AF252164

Calculus

Readings from the *Mathematics Teacher*

edited by

LOUISE S. GRINSTEIN

Kingsborough Community College
of the City University of New York
Brooklyn, New York

and

BRENDA MICHAELS

Military Sealift Command, Atlantic
Bayonne, New Jersey

National Council of Teachers of Mathematics

The special contents of this book are
Copyright © 1977 by
THE NATIONAL COUNCIL OF TEACHERS OF MATHEMATICS, INC.
1906 Association Drive, Reston, Virginia 22091
All rights reserved

Library of Congress Cataloging in Publication Data:

Main entry under title:
Calculus.

Includes bibliographies and indexes.
1. Calculus. I. Grinstein, Louise S.
II. Michaels, Brenda. III. The Mathematics
teacher.
QA303.C177   515'.08   77-6615
ISBN 0-87353-031-4

Printed in the United States of America

15.08
126n

Contents

FUNDERBURG LIBRARY
c.1
MANCHESTER COLLEGE

iii

3. Functions   45

4. Limits   93

5. Differentiation

6. Indeterminate Forms

7. Integration

8. Numerical and Mechanical Methods   171

9. Infinite Series   183

10. Special Numbers: e and π 191

Introduction

This book collects in one volume some of the many valuable suggestions that calculus instructors have had published as articles in the *Mathematics Teacher* from the time of its inception in 1908 through 1973.

We have attempted to include the widest range of subject matter while preserving balanced coverage. The principal criterion for selecting an article was whether it would be helpful to calculus teachers and students. Articles dealing exclusively with curricular structure were not included. Where several articles overlapped, we chose the most comprehensive.

Our classification has been patterned after that in *Selected Papers on Calculus* (T. M. Apostol et al. [Washington, D.C.: Mathematical Association of America, 1969; distributed by Dickenson Publishing Co.]), although modifications to the original categories have been made in order to accommodate the material in hand. The selected articles are arranged in eleven categories, some of which have been subdivided as appropriate. Articles in each subcategory are arranged chronologically, incorporating corrections of any noted typographical errors. An annotated bibliography at the end of each subcategory lists additional articles of interest from past issues of the *Mathematics Teacher*.

It can be legitimately asked, "Why another calculus book when calculus texts swamp the market?" We wished to collect in one volume some of the many valuable suggestions for more meaningful and innovative instruction that calculus instructors have developed over the years. The need for ready accessibility to such material is much more pressing today when calculus, once the province of colleges, is being taught at the high school level as well.

For the student, this book can provide deeper insight; for the instructor, it can be a source of motivational and reference material. For student and teacher alike, as well as for the concerned citizen, it can serve as a springboard for a more intensive study of the calculus. It can be a guide and a stimulus for more creative thinking about calculus in particular and mathematics in general.

1

Historical Overview

According to James E. Parker (see "The Teaching Objectives in a First Course in the Calculus"), a major objective in teaching calculus is "to give the student an understanding of the fundamental concepts of the calculus and a point of view relative to the historical background out of which these concepts grew." Current texts frequently fail in the latter regard. Most historical and biographical details concerning contributors to the growth of calculus are omitted. As a result, students often complete a calculus course with little appreciation of its historical development. The articles presented in this section were selected to provide such background material.

The first article, by Boyer, provides a chronology of calculus development as well as a detailed analysis of L'Hôpital's (or sometimes *L'Hospital*) *Analyse des infiniment petits*. This text, published in 1696, had a major influence on the further evolution of calculus. Schrader's article, on the other hand, concentrates on an in-depth analysis of the Newton-Leibniz controversy concerning the discovery of calculus. The effect of this quarrel has been far reaching. Although mathematical thought progressed in both Germany and France, English mathematicians isolated themselves for many years from the mainstream of mathematical development.

This section provides reference material that can be used to revitalize and add new dimensions to the teaching of calculus. Students should be helped to realize that the calculus of their textbooks has resulted from the arduous labors of many mathematicians over a period of centuries.

The First Calculus Textbooks

By CARL B. BOYER

Brooklyn College, Brooklyn, N. Y.

THE YEAR 1946 marks just a quarter of a millennium since the appearance of the *Analyse des infiniment petits* of L'Hospital.[1] This book, the first published text on the calculus, could boast truly that in its day it filled with distinction that ubiquitous lure of textbook writers—the long-felt need. Moreover, its influence and popularity dominated the whole of the eighteenth century, the period during which the new analysis developed until it completely overshadowed other branches of mathematics.

Historically integration antedates differentiation by some two thousand years. The ancient Greek method of exhaustion and the infinitesimal mensurations of Archimedes represent early examples of limits of integral sums; but it was not until the seventeenth century that Fermat found tangents and critical points through methods equivalent to evaluating limits of difference quotients. It was the discovery of the inverse nature of these two processes, together with the consequent exploitation of the anti-derivative in determining limits of sums, which leads one to call Newton and Leibniz the inventors of the calculus. Differentiation, both direct and inverse, became the basic algorithm in a new and powerful part of mathematics. Integration was nothing but "the memory of differentiation," and it was not until a century and a half later that some attention again was directed to the summation concept in the calculus.

The earliest expositions of the new analysis were but the barest outlines of the rules of differentiation. The calculus first appeared in print in a six-page memoir by Leibniz in the *Acta Eruditorum* of 1684.

This contained a definition of the differential and gave brief rules for determining the differentials of sums, products, quotients, power, and roots. It included also a few applications to problems of tangents and critical points. This account was so short and contained so many misprints as to be largely enigmatic to the best mathematicians of the age.[2] The basis of Newton's method of fluxions was first published unobtrusively in the form of lemmas in the *Principia* of 1687, more than a score of years after his discovery of the calculus. Here one finds some properties of limits, as well as cryptically brief directions for finding infinitely small "moments" of products, powers, and roots.[3]

The early years of the calculus resemble the infancy of analytic geometry in that little was done to popularize the subject. It is said that when Huygens about 1690 wished to master the new methods, there were not half a dozen men qualified to expound this analysis. Newton between 1669 and 1676 had composed several treatises on the elements of fluxions, but these did not begin to appear in print until 1704. John Craig in 1685 and 1693 published two works based in part on the method of Leibniz, but these were not intended as introductions to the subject and were moreover difficult to read because of the geometrical form in which they were cast. On the Continent the Bernoulli brothers with some effort achieved

---

[1] G. F. A. de L'Hospital, *Analyse des infiniment petits, pour l'intelligence des lignes courbes*, Paris, 1696. Page references indicated in this paper refer to the original edition.

[2] See Gustav Eneström, "Über die erste Aufnahme der Leibnizschen Differentialrechnung," *Bibliotheca Mathematica* (3), IX (1908–1909), 309–320. An English translation of Leibniz's paper of 1684 is found in D. E. Smith, *A Source Book in Mathematics* (New York, 1929), pp. 619–626.

[3] Sir Isaac Newton, *Opera quae exstant omnia* (ed. by Samuel Horsley, 5 vols., Londini, 1779–1785), I, 251, II, 277–280.

ANALYSE

DES

INFINIMENT PETITS,

Pour l'intelligence des lignes courbes.

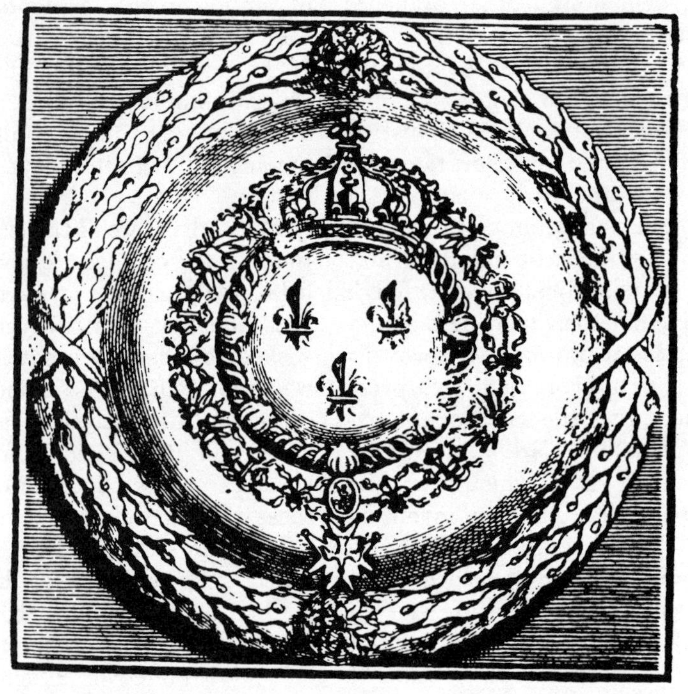

A PARIS,

DE L'IMPRIMERIE ROYALE.

M. DC. XCVI.

TITLE-PAGE OF THE FIRST EDITION OF L'HOSPITAL'S TEXTBOOK
(Courtesy of Columbia University Library)

MARQUIS DE L'HOSPITAL
(Courtesy of David Eugene Smith Library)

a thorough understanding of the subject, and it was Jean Bernoulli who then instructed L'Hospital, while the latter in turn passed his knowledge on to Huygens. That there was an atmosphere of enthusiasm surrounding the calculus toward the close of the seventeenth century is to be ascribed in large part to the research articles published in the learned periodicals of their day by Leibniz, L'Hospital and the Bernoullis. The large mathematical public had, before 1696, no easy introduction to the subject. Jean Bernoulli about 1691–1692 had composed two little textbooks, but their publication was long delayed: that on the integral calculus finally appeared in 1742,[4] and that on the differential calculus almost two hundred years later still.[5] That the text of L'Hospital achieved such prompt success is ascribable largely to the fact that it was the first book to supply what the many had long awaited—an elementary introduction to a new and fascinating branch of mathematics.

Guillaume-François-Antoine de L'Hospital, Marquis de St. Mesme,[6] was born in 1661. His interest in geometry was aroused at an early age, and by the time he was fifteen years old he solved difficult problems on the cycloid which Pascal had proposed. He became a captain of cavalry, but later gave up army life to devote more time to the study of mathematics. When Jean Bernoulli was in Paris in 1692, L'Hospital for four months studied the new infinitesimal geometry under him.

From that time on he joined the ranks of the mathematical elite and became the chief exponent of the calculus in France. Recognizing the lack of elementary expository works, he wrote to Bernoulli in 1695 that he was on the point of publishing a work on conic sections and that he proposed to add a little treatise on the differential calculus. L'Hospital indicated that he would render his master the justice he deserved, and he added modestly that the projected work would serve merely as an introduction to a more elaborate treatise, *De scientia infiniti*, which Leibniz planned to write.[7]

The promised book on conics was delayed by the author's illness, and appeared posthumously in 1707; but the *Analyse des infiniment petits* was published promptly in 1696. In the preface L'Hospital points out that he owed much to Leibniz and the Bernoullis, especially "the young professor at Groningen"; and he says that they can take whatever credit they wish, leaving to him what they will.[8] After Jean Bernoulli had received a copy of the book, he wrote and thanked l'Hospital for mentioning him in the volume, and he promised to return the compliment when he in turn should have composed something. He wrote that the work was admirably done, and praised it for the sound arrangement of propositions and for the intelligibility of the exposition. To L'Hospital's suggestion that he should write an integral calculus as a sequel to the *Analyse*, in view of the protracted delay on the part of Leibniz, Bernoulli replied that he was so distraught by domestic embarrassments that he could not carry out the project.[9]

Seldom has a textbook in mathematics been received by savants with such eagerness as was that of L'Hospital. It opened

---

[4] Jean Bernoulli, *Opera omnia* (4 vols., Lausanne and Geneva, 1742), vol. III; also *Die erste Integralrechnung*. Eine Auswahl aus Johann Bernoullis mathematischen Vorlesungen über die Methode der Integrale und anderes aufgeschrieben zum Gebrauch des Herrn Marquis de l'Hospital in den Jahren 1691 und 1692. Translated from the Latin, Leipzig and Berlin, 1914.

[5] Jean Bernoulli, *Die Differentialrechnung aus dem Jahre 1691/92*, Ostwald's klassiker, No. 211, Leipzig, 1924.

[6] A satisfactory biographical account is found in J. F. Michaud, *Biographie Universelle* (2nd ed., Paris, 1880), XXIII, 448–451. In later years the family name came to be spelled L'Hôpital instead of L'Hospital.

[7] See L'Hospital, *op. cit.*, preface.

[8] *Ibid.*

[9] Eneström, "Sur le part de Jean Bernoulli dans la publication de l'Analyse des infinment petits," *Bibliotheca Mathematica* (new series), VIII (1894), 65–72. Cf. O. J. Rebel, *Der Briefwechsel zwischen Johann* (I). *Bernoulli und dem Marquis de l'Hospital*, Bottrop i.w., 1934.

the door to a powerful tool which previously had been accessible only to the very gifted, and thus made it possible for the average reader to solve problems which had defied the best efforts of the geometers of Greece. But the reception was not one of unmixed acclaim. Echoing the contemporary literary controversy on "ancients vs. moderns," some admirers of the classic synthetic geometry cast doubt on the soundness of the new analytic approach. The Académie des Sciences in 1701 was sharply divided over the *Infiniment petits*. The Abbé Gallois led a polemic attack on the methods of the book, ably supported by Rolle.[10] L'Hospital, although he was a member of the Académie, did not reply; but his cause was warmly and effectively sustained by Varignon, Saurin, and others. The controversy reached such a pitch that the Académie finally felt obliged to name a commission to judge the question. Interest in the attack now began to wane, but L'Hospital did not witness the ultimate triumph of the infinitely small, for he died in 1704.

Jean Bernoulli meanwhile had observed with apparent jealousy the growing success of his protégé's book. During the years immediately following the author's death, he wrote letters attacking the work and practically accusing L'Hospital of plagiarism. L'Hospital's general prefatory acknowledgment, Bernoulli argued, did not do him justice inasmuch as he was in effect the true author of almost all the substance of the book. Historians of mathematics have been divided in judgment on this matter, but a thorough analysis of the question made by Eneström reaches the conclusion that the broad claims of Bernoulli with respect to the authorship of the material are not substantiated.[11]

[10] Académie des Sciences, *Histoire et mémoires*, 1701 (Amsterdam, 1735), Histoire, pp. 114 ff. This controversy was the second of three important attacks on the calculus. The first was that of Nieuwentijdt against the work of Leibniz in 1695–1696. The third was the famous *Analyst* controversy of 1734–1735 concerning the calculus of Newton.

[11] Eneström, *op. cit.*

The *Analyse des infiniment petits* represents the first systematic treatment of the calculus, and as such it presents a good picture of the level of the subject at that time. The preface includes a brief historical account in which the author, while admitting that "Newton also had a kind of calculus," prefers and emphasizes the contributions of Leibniz. The treatise itself is divided into ten chapters or sections, of which the first (pp. 1–10) is devoted to the fundamental definitions, assumptions, and rules of procedure. A variable quantity is defined loosely as one which continually increases or diminishes; and the infinitely small portion by which a variable quantity changes is called the "difference" or differential. Two postulates are laid down: that one can take as equal two quantities which differ only by an infinitely small quantity; and that a curved line can be considered as an infinite assemblage of infinitely small straight lines which determine, by the angles which they make with each other, the curvature of the curved line. To L'Hospital these premises appeared to be "so self-evident as not to leave the least scruple about their truth and certainty on the mind of an attentive reader," but he adds that if space permitted he could prove them in the manner of the ancients. The crudeness of these fundamental principles is an indication of the infancy of the subject rather than of carelessness on the part of the author. The language of L'Hospital is eminently restrained in comparison with that of contemporaries who wrote such things as

$$\frac{d^3y}{d^2x} = d^3y \int^2 x \ \text{ or } \ \infty \cdot \infty^{\infty-1} = \infty^{\infty}.$$

The basic differential formulas for algebraic functions—sums, products, quotients, powers, and roots—are derived by L'Hospital in the customary Leibnizian manner, infinitesimals of higher order being neglected. Independent and dependent variables were not always clearly distinguished at the time, and here L'Hospital took a step in advance by stating formally the essential implicit function formula in the case of powers of expressions involving one or more variables. A modern reader will be struck by the absence of rules for the differentiation of transcendental functions. These were known at the time, but they had not been popularly formalized. L'Hospital had suggested to Bernoulli that he might wish to append to the *Analyse* his rules for the calculus of logarithms, but this was not done. However, the equivalent of these was implied by the well-known results on the area between the hyperbola and its asymptotes. The differentials of the trigonometric functions were at the time likewise geometrically inferred from the application of the characteristic triangle to the circle. The inverses offered no difficulty in view of the known formula for implicit functions.

The second section of the *Infiniment petits* (pp. 11–40) is devoted to problems of tangency. Inasmuch as the tangent is defined as the prolongation of an infinitely small side of the polygon which makes up the curved line, one sees immediately from the similarity of triangles that the subtangent is given by $y(dx/dy)$. Through this relationship, the author finds the tangents to the curves of his day: the higher hyperbolas and parabolas, the spirals of various kinds, the quadratrix of Dinostratus, the conchoid of Nicomedes, the cissoid of Diocles, the cycloid, the tractrix, and the curve of logarithms. Some of these curves are defined analytically, while others are described kinematically or in terms of geometrical properties. L'Hospital follows the Cartesian classification of curves into geometrical and mechanical, but inasmuch as no special notations are used for transcendental functions, only in the former category is "the nature of a curve" given by an ordinary equation.

The third chapter (pp. 41–54), on maxima and minima, opens with a definition of the basic terms: If the ordinate (appliquée) increases, as the abscissa (coupée) increases to a given value E, and thereafter decreases, then the ordinate at E is called the greatest; and similarly, *mutatis mutandis*, for the least ordinate. The treatment here differs from that in modern textbooks not only in being less rigorous, but also in that cusp maxima and minima are treated as extensively as extrema having horizontal tangents. L'Hospital points out that the differential of an increasing quantity is positive, that of a decreasing quantity is negative. Moreover, no quantity which varies continuously can go from positive to negative without going through zero or infinity. Hence to find maxima and minima of an algebraic expression, L'Hospital equates to zero both the numerator and the denominator of the differential, discarding results which lead to a contradiction. A few cases of applications in geometry and science are included. He does not explicitly state the various tests distinguishing maxima from minima, but these were well-known to the author and his times.

Points of inflection and of *rebroussement* (cusp points for which the tangents are horizontal) are treated in section four (pp. 55–70), and here the thorny question of second order differences arose. Leibniz had been unable to give a satisfactory definition of differentials of order greater than one, and the first attack on the calculus, launched by Nieuwentijdt in 1695–1696 while L'Hospital was preparing the *Analyse*, was centered on this inadequacy. In his preface (p. eiij) L'Hospital claimed that Leibniz had satisfactorily answered Nieuwentijdt's doubts about the existence of higher differentials, but this is belied by the author's own definition: "The infinitely small portion by which the difference of a variable quantity increases or decreases continually is called the difference of the difference of this quantity, or its second difference." That he obtained correct results in this connection is due less to his definition than to the pragmatic rule which follows it: "One takes as constant whichever difference one wishes, and treats the others as variable quantities." For example, the differential of the function xy being $xdy+ydx$, the second differential is either $dd(xy)=xddy+2dxdy$ or $dd(xy)=yddx+2dxdy$. Here one sees the need to distinguish between the differentials of independent and dependent variables. This distinction was not at the time sufficiently emphasized, and even to-day numerous textbooks founder on the differential through neglect in this connection. L'Hospital then gives the geometrical description of points of inflection and *rebroussement* in terms of convexity and concavity, and he points out that these points are determined by making the second differential either zero or infinitely great. A more rigorous treatment is not, of course, to be expected of that period.

The concept of radius of curvature as such is not presented in the modern manner but is given in the context in which it arose historically. Huygens had been led by his pendulum clock to study the subject of evolutes and involutes (*developpées* and *developpants*), and this material is presented by L'Hospital in section five (pp. 71–103). Toward the close of this chapter, the longest of all, the author points out in a casual sort of way that $\overline{dx^2+dy^2}^{\frac{3}{2}}/-dxddy$ is a general expression for "the radius of an evolute." This is, of course, the equivalent of the modern formula for radius of curvature, where y is a function of x.

Sections six and seven (pp. 104–119, 120–130) are on caustics by reflection and refraction. These are not now traditional topics in elementary calculus, but they were popular new topics in the late seventeenth century. Moreover, L'Hospital and his master, Bernoulli, had shared with Tschirnhaus in their development. Section eight (pp. 131–144) is on envelopes of families of straight lines, and includes the well-known method of Leibniz of differentiating with respect to the parameter.

The portion of the book which in its day roused most discussion was section nine (pp. 145–163). This chapter carries the unenlightening heading, "Solution of some problems depending on the preced-ing methods," but it involves what would now be called indeterminate forms. The presentation is largely geometrical, but the basic result is the so-called "rule of L'Hospital." To find "the value" of a rational expression in x which for the abscissa in question takes the form 0/0, he determines the ratio of the "differences" of the numerator and denominator for this abscissa. This rule appears not to have been familiar at the time, and so the author here might well have indicated the source of his material. The author never claimed the rule as his own, but when his friend Saurin implied that it was due to Leibniz, protest was raised by Jean Bernoulli that here, as elsewhere, L'Hospital had not given him due credit. Bernoulli's claim to the rule seems to be substantiated, and his name, rather than L'Hospital's, should be applied to it. But it should be pointed out that the traditional nomenclature is due to the fickleness of fortune rather than to any ill intent on the part of L'Hospital. The *Analyse des infiniment petits* occupies a place in the calculus somewhat analogous to that of Lavoisier's *Traité élémentaire* in chemistry almost a century later. Both books would have been more attractive had they included a substantial historical background, but in neither case should the omission of this be interpreted as an argument *e silencio* that the author claims credit for more than the general arrangement and exposition of the subject matter. One does not demand of writers of textbooks the same scrupulous regard for the citation of sources that one expects in the publication of the results of research work.

The tenth and last section of L'Hospital's treatise (pp. 164–181) is somewhat out of harmony with the remainder. It is a clever bit of salesmanship in which the author contrasts the elegant methods of the calculus with the awkward anticipatory procedures of Descartes and Hudde for determining maxima and minima. On the basis of this comparison L'Hospital emphasizes that the method of

Leibniz "gives general solutions, whereas the other furnishes only particular ones, and it extends to transcendental curves and it is not at all necessary to avoid incommensurables, which often is impracticable." This closing sentence of the first textbook on the calculus reminds one of the title of Leibniz's first paper on the new analysis a dozen years before: "A new method for maxima and minima, as well as tangents, which is not impeded by irrational quantities."

The *Analyse* of L'Hospital was composed two hundred and fifty years ago, and hence the material it contains does not in all cases parallel that which appears in modern textbooks. One fails to find, for example, the notions of function and limit. The first of these was in the air in L'Hospital's day, but it was popularized half a century later in Euler's *Introductio ad analysin infinitorum;* the latter was likewise known in a general sense through the calculus of Newton, but it was a hundred years before it became basic in the treatises of Cauchy. One misses also the now traditional work on Taylor's series. This, too, was known to James Gregory and Jean Bernoulli, but it was formalized after L'Hospital's time through Brook Taylor's work on finite and infinitesimal differences. Curve tracing is a further topic which is lacking in the *Analyse*. The seventeenth century here invented the analytical tools, but it was the eighteenth century which first applied them to the systematic study of a wide variety of curves. It is strange to relate also that L'Hospital's text is deficient in applications to science. Seventeenth century dynamics afforded a golden opportunity for apt and timely illustrations of the new calculus, but we are told in the preface (p. eij) that illness prevented the author from fulfilling his intention to add a section on applications to physics. The language and notation of the *Analyse* differ slightly from that now in use; there is an anachronistic emphasis on the ideas of ratio and proportion; and the material

is divided into propositions, corollaries, and scholia. Nevertheless, a modern reader should have no difficulty in following the author's exposition.

The *Analyse des infiniment petits* appeared in a second edition at Paris in 1715. This was in reality a reissue of the original, but it contained numerous typographical errors. The work was reprinted the following year, and again in 1720. By this time the popularity of the calculus was such that a textbook on a lower level was desired. Publishers sought to capitalize on this desire and on the popularity of L'Hospital's work by providing commentaries. In 1721 Crousaz issued at Paris his *Commentaire sur l'analyse des infiniment petits,* consisting of an elementary introduction and a series of supplementary notes. This book was damned by faint praise on the part of Jean Bernoulli and met with severe criticism from Saurin. Four years later there appeared a posthumous and successful supplementary volume by Varignon, *Eclaircissemens sur l'analyse des infiniment petits.* The author appears to have had in mind the preparation of a new edition of L'Hospital's work, together with added notes. The publisher, however, felt that inasmuch as L'Hospital's *Analyse* already "had appeared several times in France and in foreign countries," and since "most men in the profession already had copies," he could save the public some expense by issuing Varignon's *Eclaircissemens* as a separate volume rather than as part of a new edition. This volume is on a much higher level than that of Crousaz, and although it follows L'Hospital section by section, it adds to this fresh problems and methods resulting from later research.

The rift between British and Continental mathematicians during the eighteenth century was not sufficiently wide to prevent Stone in 1730 from publishing an English translation of L'Hospital's *Infiniment petits,* ten years before Buffon reciprocally prepared a French translation of Newton's *Method of Fluxions.* The full title of Stone's book is, *The method of fluxions both direct and inverse. The former being a translation from the celebrated Marquis de L'Hospital's Analyse des infiniments petits: and the latter supply'd by the translator.* The translator opens with praise of the author's work, "the Character whereof is so well establish'd," but adds that out of regard to Sir Isaac Newton he has altered some of the language and notation of the original to conform to the Newtonian. Increments or differentials, for example, are taken as fluxions, inasmuch as they are proportional to the latter; and these are denoted by $\dot{x}$, $\ddot{x}$, $\dddot{x}$, etc., instead of by dx, ddx, $dddx$, etc. A few years later Bishop Berkeley took full advantage, in the famous *Analyst* controversy of the confusion then existing in Britain between fluxions and infinitely small quantities.

Stone's translation, in part I, does not depart much from the original, and the diagrams are faithful reproductions of those of L'Hospital. Some new material is added, however, on logarithms, exponentials, and trigonometric functions. The second part or appendix (separately paginated) is on the inverse method of fluxions, and this constitutes a new work. L'Hospital apparently had hoped that his treatise would sometime be completed by a second part on the integral calculus. The pages of the *Analyse* (even in later editions) carry the heading *I. Part.*, but the author never completed another *partie*. Stone was but one of several who undertook to compose a suitable second half. His attempt was so successful that it was translated into French and published in 1735, "servant de suite aux infiniment petits de M. le Marquis de l'Hôpital."

The *Analyse des infiniment petits* continued throughout the eighteenth century to be the standard elementary manual on higher mathematics.[12] During the latter half of the century it found a worthy rival in the *Instituzioni analitiche* of Maria Agnesi (Milan, 1748), but frequency of publication indicates that the vogue for L'Hospital's work continued. It appeared in new French editions, revised and enlarged, in 1758 (Paris), 1768 (Avignon), and 1781 (Paris), and in Latin translations in 1764 and 1790 (both at Vienne).[13]

Toward the turn of the century, however, Lacroix[14] began the publication of his renowned series of textbooks, and one of these, the *Traité élémentaire de calcul différentiel et de calcul intégral*, (Paris, 1802) took the place during the nineteenth century which L'Hospital had held a hundred years earlier. The concepts of function and limit now took over the field which the infinitesimal "ghosts of departed quantities" had so long dominated. But the work of the first textbook on the calculus had been well done. The enthusiasm for the subject which it kindled and nourished has never abated, and modern textbooks are but new monuments to the influence of the *Analyse des infiniment petits*.

---

[12] The Abbé Sauri in his *Cours complèt de mathématiques* (5 vols., Paris, 1774), III, xij, refers to L'Hospital's *Analyse* as "connue de **tout le mond**."

[13] This paper does not pretend to give a complete listing of editions. I have examined those of 1696, 1715, 1730, 1768, and 1781, and the commentaries by Crousaz and Varignon, all of which are available at the New York Public Library or at Columbia University. Editions other than these have been listed on the basis of references found in standard sources, such as Poggendorff and the catalogues of the British Museum and the Bibliothèque Nationale. It is of interest to note that L'Hospital's *Traité analytique des sections coniques* likewise enjoyed a considerable popularity during the eighteenth century, appearing in editions of 1707, 1720, 1723 (English translation by Stone), 1770 (Venice), and 1776.

[14] S. F. Lacroix was perhaps the most prolific textbook writer of modern times, if allowance is made for multiple editions. In 1848 there appeared at Paris the 20th edition of his *Traité élémentaire d'arithmétique* and the 16th edition of his *Élémens de géométrie*. The 20th edition of his *Élémens d'algèbre* was published at Paris in 1858, and the ninth edition of the *Traité élémentaire de calcul* in 1881. And the above figures do not include the large number of editions in other languages.

The Newton-Leibniz controversy concerning the discovery of the calculus

by Dorothy V. Schrader, Southern Connecticut State College,
New Haven, Connecticut

THE STATE OF ANALYSIS IN THE SEVENTEENTH CENTURY

The seventeenth century was one of activity and advancement in the world of mathematics. Analytic methods had become familiar tools to most of the mathematicians of the period; geometry was being employed to verify and demonstrate analytic conclusions; and special attention was focused on problems dealing with the infinite.

> The time was indeed ripe, in the second half of the seventeenth century, for someone to organize the views, methods, and discoveries involved in the infinitesimal analysis into a new subject characterized by a distinctive method of procedure.[1]*

Unfortunately, not one but two men did just that. We say unfortunately, because the methods of the calculus developed by Sir Isaac Newton in England and Gottfried Wilhelm Leibniz on the continent were essentially the same, yet the dispute over the rights of the two discoverers developed into a controversy which has not yet been settled. Both these mathematicians and their followers stooped to tactics which were most unworthy of men of intelligence and honor; as a result, the development of mathematics in England was brought almost to a standstill for a full century. The judgment of history seems to be that credit belongs to both individuals equally.

> [I]t might be far better to speak of the evolution of the calculus. Nevertheless, inasmuch as Newton and Leibniz, apparently independently, invented algorithmic procedures which were universally applicable and which were essentially the same as those employed at the present time in the calculus, . . . there will be no inconsistency involved in thinking of these two men as the inventors of the subject.[2]

It must be remembered, however, that these two inventors are not responsible for the definitions and ideas which are considered basic to the calculus today. Only in the present generation, after more than two centuries of development, has there been laid the foundation of mathematical rigor on which the calculus now rests.

NEWTON'S METHOD OF FLUXIONS

Newton called his discovery the Method of Fluxions and described it in terms of geometry. The direct method of fluxions can be summarized in the solution of the mechanical problem: "The length of the Space described being continually given, to find the Velocity of the Motion at any time proposed,"[3] or "The relation of the flowing Quantities being given, to determine the relation of their Fluxions."[4] Else-

* Footnote references are to be found at the end of this article.

where, Newton refers to these flowing quantities as Fluents. The inverse method of fluxions can be summarized in the inverse of the problem given above: "The Velocity of the Motion being continually given, to find the Length of the Space described at any Time proposed,"[5] or "An Equation being proposed including the Fluxions of Quantities, to find the Relation of those Quantities to one another."[6] These direct and inverse methods of fluxions are, of course, the familiar differential and integral calculus. Newton further described his fluents as quantities which are to be considered as gradually and indefinitely increasing; these he represented by the last letters of the alphabet: x, y, z. The velocities by which every fluent is increased by its generating motion, he called fluxions and designated by "pointed" or "prickt" letters, corresponding to the fluents involved: $\dot{x}$, $\dot{y}$, $\dot{z}$. The moment of a fluent, its velocity multiplied by an infinitely small quantity, o, he represented in the fluxional notation as $\dot{x}o$.[7]

Leibniz' differential and integral calculus

Instead of the flowing quantities and velocities of Newton, Leibniz worked with infinitely small differences and sums. He used the now familiar $\dfrac{dy}{dx}$ instead of Newton's dotted letters for the derivative symbol; he used $\int$ for his integration symbol while Newton used either words or a rectangle enclosing the function. Newton himself asserted that his "prickt" letters were equivalent to Leibniz' $\dfrac{dy}{dx}$ and that by $\boxed{\dfrac{aa}{64x}}$ he meant the same thing that Leibniz meant by $\displaystyle\int \dfrac{aa}{64x}$.[8]

Leibniz was more interested in developing a notation for his new method

than was Newton. The Englishman used the dot symbolism in his fluxionary calculus but did not employ it in his treatise on analysis nor in his famous *Principia*. In one report, printed anonymously but commonly believed to be written by Newton, we read, "Mr. Newton doth not place his Method in Forms of Symbols, nor confine himself to any particular Sort of symbols for Fluents and Fluxions."[9] This independence of symbols, which Newton apparently thought praiseworthy, is today considered one of the major weaknesses of his method. Modern mathematics is almost wholly dependent upon the symbols by which it is expressed, so much so that someone has characterized mathematics as the science in which one operates with and on symbols, neither knowing nor caring what these symbols mean, if indeed they have any meaning. Felix Klein has lauded Leibniz for that very type of symbolism. He notes that $\int y$ and not $\int y\, dy$ was used in Leibniz' manuscripts, and credits him with being the founder of modern, formal mathematics for recognizing that it makes no difference what, if any, meaning is attached to the differentials, but that, if appropriate rules of operation are defined for them and the rules correctly applied, something reasonable and correct will result.[10]

Whether or not Klein errs in attributing too deep a perception to Leibniz is difficult to determine, but it is true that Leibniz was much concerned with finding the best possible notation for his calculus. He experimented with various symbols, explained them to different people, asked advice of a number of mathematicians, and used the dx and dy notation for a long time before he published it. He finally selected the particular form of notation because he saw the great need of being able to identify easily the variable and its differential.[11] Johann Bernouilli, who worked with him on integration, wanted to call the new branch of mathematics "calculus integralis" and use I as an integration symbol; Leibniz preferred the name "calculus summatorius" and the symbol $\int$. Today, we use Bernouilli's name and Leibniz' symbol.

The method of fluxions and the differential and integral calculus differ in more than their notation. While the two methods are essentially the same in that they can be reduced to a common method, they start from different principles. Newton made use of infinitely small quantities to find time derivatives, which he called fluxions; he stated specifically that his mathematical quantities were to be considered as described by continuous motion, not as existing in infinitely small parts. Leibniz, on the other hand, made the infinitely small quantities themselves the basic concepts in his differentials.[12] Newton dealt with a finite quantity which is the ratio of two infinitely small quantities, the ratio of velocities; Leibniz dealt with the finite sum of an infinite number of infinitely small quantities.[13]

The quarrel

The famed Newton-Leibniz controversy concerning the discovery of the calculus involves more than merely the question of priority of time. Mutual accusations of plagiarism, secrecy which manifested itself in cryptograms, letters published anonymously, treatises withheld from publication, assertions of friends and supporters of the two men, national jealousies, and the efforts of would-be peacemakers, all serve to complicate the situation and make such a tangle of truth and falsehood, information and misinformation, that it can probably never be solved conclusively. We can at best review the major events and draw a few general conclusions.

In June, 1669, Newton, who had become interested in mathematical analysis during his days as an undergraduate at Cambridge, sent to Isaac Barrow, the geometer, a manuscript entitled "Analysis per Equationes Numero Terminorum Infinitas," in which the underlying prin-

ciples of the theory of fluxions were indicated. Barrow was impressed with the work, and, in a letter dated June 20, 1669, mentioned it to a mathematician friend, Collins. On July 31 of the same year, he sent the manuscript to Collins, who copied it and returned the original to Barrow. Collins was in correspondence with many of the leading mathematicians of England and the continent; in letters dated from 1669 to 1672, he communicated Newton's discoveries to Gregory, Bertet, Vernon, Slusius, Borelli, Strode, Townsend, and Oldenburg.[14]

At about the same time, Leibniz was working on problems suggested by the theories of Cavalieri; in 1671, he dedicated to the French Academy a paper, "Theoria Motus Abstracti," in which he showed that he was considering the use of infinitely small quantities in these problems.[15] This was not a well-developed theory of differential calculus, but it does indicate that Leibniz' mind was working along the lines of infinitesimal analysis, as was Newton's.

In 1672, Newton composed a treatise on fluxions, which, however, was not published until John Colson translated it from Latin into English and published it in London in 1736 under the title, *Method of Fluxions*. Why it was not published at the time it was written seems not to be known.

Early in 1673, Leibniz was in London where he visited Oldenburg, a fellow-countryman and the secretary of the Royal Society. As he attended meetings of the society and read mathematical papers before it, it is quite possible that he met Collins, who was a friend of Oldenburg. In 1890, notes of this London visit were discovered in the royal library at Hanover, where Leibniz had been librarian. These notes show extracts from Newton's work on optics, which was not published until 1704, but make no mention of mathematics. Hathaway[16] finds this omission very strange and somehow indicative that Leibniz saw Newton's paper, "De Analysis," during this visit.

By March of 1673, Leibniz was back in Paris, where he began a serious study of geometry with Huygens. In July of the same year, he wrote to Oldenburg, discussing his work on series. In reply, Oldenburg told him of some of Newton's and Gregory's discoveries on series and tangents. Leibniz, working with infinitely small sums and differences, defined the general problem of the area of the curve and from it developed the algorithm of the differential and the integral calculus, a logical outcome of the studies on which he had been engaged for several years. Manuscripts in the library at Hanover show that by the end of 1675 he had a clear idea of the principles of the calculus and had invented the notation.[17]

On June 13, 1671, Newton, answering a request from Leibniz, wrote a brief account of his method of quadrature by means of infinite series and discussed the binomial theorem. Leibniz replied, August 27, 1676, and asked for more details. In September of that year, Leibniz was again in London, where he spent a week with Collins, saw Newton's manuscript of the "De Analysis," and made copious notes from it. One author asserts that "since its pages were open freely to him at that time it is constructive proof that they were as freely open to him for the two months in 1673 that he was in London."[18] This reasoning seems obscure, but it may be that Hathaway's testimony is colored by anti-German feeling,[19] which is perhaps understandable, as he wrote shortly after World War I.

After his return to Paris, Leibniz received another letter from Newton, dated October 24 and sent through Oldenburg. This was fifteen closely written pages, discussing series and mentioning fluxions, but giving no detailed information. Newton himself said later that he had told Leibniz of his method of fluxions but disguised it in an anagram of transposed characters under the sentence, "Data aequatione quotcunque fluentes quantitates involvente, fluxiones invenire, et

vice versa," which, transliterated, reads, "Given any equation whatsoever involving flowing quantities, to find the fluxions, and vice versa." This as an anagram would be: 6a 2c d ae 13e 2f 7i 3l 9n 4o 4q 2r 4s 9t 12v x. One wonders how much Leibniz could ever make out of that anagram, and having reconstructed Newton's sentence, could he deduce the method of fluxions from it? "Whoever can form a certain sentence properly out of 6 a's, 2 c's, a d and so on, will see as much as one sentence can show about Newton's mode of proceeding."[20]

The question immediately arises as to why Newton bothered to tell Leibniz anything at all if he was going to conceal it so thoroughly in his anagram. The device is not unprecedented. Galileo used to give his discoveries to his friends in the form of carefully dated cryptograms in order to establish priority. Sometimes academies and learned societies were the trustees for intellectual secrets. Such secrecy was considered necessary to protect the rights of the inventors. There was much jealousy among the learned who, in their desire to conceal their discoveries, were wont to publish theorems without proof or demonstration. Newton deposited his method, by anagram, in the hands of his rival.[21]

There is another question concerned with this famous anagram. If, as has been frequently stated, Newton had given a clear indication of his method of fluxions in his "De Analysis" of 1669, which was, with his knowledge, being circulated and discussed by Collins, why did he consider it necessary seven years later to conceal the method? Could it be that he had merely hinted at it in 1669 and did not have it so well-developed as has been supposed?

After receiving Newton's letter, Leibniz replied on June 21, 1677, through Oldenburg, that he too had a method of drawing tangents, not by fluxions but by differentials; he quite frankly and openly explained the differential calculus and its applications. However, he did not tell Newton that he had seen the 1669 manuscript in London but a few weeks before, that he knew of Newton's work, and that the anagram was useless. Was Leibniz being unfair to Newton or was the 1669 treatise less complete than it is reputed to have been? Newton did not answer and the correspondence ceased, perhaps due to the death of Oldenburg which occurred in August of 1678.

In 1683 Collins died; in 1684, in the *Acta Eruditorum* of Leipzig, Leibniz published his first paper on the calculus, an account similar to that which he had given Newton. If he had obtained his initial ideas from Newton via Collins, that might explain his desire to conceal the theft from Collins by not publishing the ideas as his own until after the death of Collins. Or is there another reason for the long delay in publication? Leibniz made a vague reference to Newton's having a method similar to his own but he made no claim to being the first or sole inventor. The first "Nova Methodus pro Maximis et Minimis" merely developed the rules for differentiation. Later works in the same publication gave an exposition of the principles from a formal viewpoint.[22]

Leibniz and the two Bernouillis were making rapid progress with the new and powerful analytic method when the first edition of Newton's *Principia* was published in 1687. In Book II, Lemma II, Newton explained the fundamental principle of the fluxionary calculus and in a scholium added:

In letters which went between me and that most excellent geometer, G. W. Leibniz, ten years ago, when I signified that I was in the knowledge of a method of determining maxima and minima, of drawing tangents, and the like, and when I concealed it in transposed letters involving this sentence (Data aequatione quotcunque, fluentes quantitates involunte, fluxiones invenire, et vice versa; that is, Having given any equation involving ever so many flowing quantities, to find the fluxions, and vice versa) that most distinguished man wrote back that he had also fallen on a method, which hardly differed from mine, except in his forms of words and symbols.[23]

Later, when the controversy was at its height, this scholium was quoted as evidence that Newton recognized Leibniz' rights as a second or simultaneous inventor. Such is not entirely the case; all Newton admitted was that Leibniz did have a method, however he learned it. Newton, after Leibniz' death, asserted that the scholium had been intended as a challenge to Leibniz to prove his priority if he could, not as an admission of his equality. In the second edition of the *Principia*, Newton added a phrase to the scholium, making it a bit more accurate. In that edition, the last words read, " . . . which hardly differed from mine except in the forms of words and symbols, and the concept of the generation of quantities."[24] In the third edition, the scholium was changed entirely and another subject inserted; neither Leibniz' name nor his work was mentioned.

In 1695, Dr. John Wallis chided Newton for being in possession of the method of fluxions for thirty years and never publishing it in its entirety; he had made reference to it in his own complete works, published in 1693, but felt that his treatment of the subject was most inadequate.

In the same year, John Bernouilli challenged Europe with two problems, to be solved in six months. To allow the mathematicians of the world time to work on these problems, Leibniz requested an additional year, which was granted. During this extension, Newton heard of the problems and solved them both in a single evening, sending his solutions to the president of the Royal Society the day after he had received the problems. Acknowledging the receipt of Newton's solutions, Leibniz, in the Leipzig *Acts*, managed to convey the impression that Newton was a pupil of his who, because he had mastered the calculus, was able to solve the problems.[25]

It was four years later, in 1699, that the hidden rivalry flared into open hostility. Fatio de Duillier, a Swiss mathematician from Geneva, who had been living in

England for about ten years and was a close friend of Newton, published a memoir in which he claimed for himself independent invention of the calculus (which claim seems to have been completely ignored) and implicitly accused Leibniz of plagiarizing from Newton.

I am bound to acknowledge that Newton was the first, and by many years the first, inventor of this calculus: from whom, whether Leibniz, the second inventor, borrowed anything, I prefer that the decision should lie, not with me, but with others who have had sight of the papers of Newton, and other additions to this same manuscript.[26]

Leibniz answered the charge, referred to Newton's scholium in the *Principia,* and, ignoring the question of priority, insisted upon his right to credit for the invention of the differential calculus. Newton ignored the entire situation and the Leipzig *Acts* refused to print De Duillier's reply to Leibniz.

Here the matter rested until Newton's *Opticks* came out in 1704. Published with the text on optics was a short treatise explaining the method of fluxions and commenting that, since theorems from the 1669 manuscript were appearing in various guises, it seemed best to the author to make the method public now. The next January, the *Acta Eruditorum* of Leipzig carried an anonymous review, later shown to have been written by Leibniz, stating:

[t]he elements of this calculus have been given to the public by its inventor, Dr. Gottfried Wilhelm Leibniz in these *Acts.* . . . Instead of the Leibnizian differences, then, Dr. Newton employs and always has employed, fluxions, which are very much the same as the augments of fluents produced in the least intervals of time, and these fluxions he has used elegantly in his *Mathematical Principles of Natural Philosophy* and in other later publications, just as Honoratus Fabri, in his *Synopsis of Geometry* substituted progressive methods for the method of Cavalieri.[27]

This innocent-sounding review provoked a storm of opposition in England. Far from being a compliment to the acuteness of the Englishman, it was, by the comparison with Fabri, a scarcely veiled attack upon Newton's integrity. Fabri had been a notorious plagiarist, discredited and

dishonored in the mathematical world for his theft of the ideas of another.[28] At the time, Leibniz denied authorship of the review, although later[29] he tacitly admitted it and gave his approval. While this attack may not have been entirely unprovoked, it most certainly was cowardly and unworthy of a man of Leibniz' standing.

John Keill, a friend of Newton, in a letter to Edmund Halley, on the laws of centripetal force, directly and openly accused Leibniz of plagiarism.

All these laws follow from that very celebrated arithmetic of fluxions which, without any doubt, Dr. Newton invented first, as can readily be proved by anyone who reads the letters about it published by Wallis; yet the same arithmetic afterwards, under a changed name and method of notation, was published by Dr. Leibniz in *Acta Eruditorum.*[30]

When Leibniz received his copy of *Transactions,* in which the letter was published (Volume XXVI, #317) he wrote to Sloane, the secretary of the Royal Society, demanding that Keill retract his accusation; he added that he felt sure that Keill had acted from rashness and not from improper motives, and that he would not consider the attack a matter of calumny. When the letter was read to the Royal Society, Newton, who was then the president, expressed displeasure at Keill's action. Keill justified himself by producing the review of Newton's *Opticks* in the Leipzig *Acts,* of which Newton had apparently been unaware until this time. Keill wrote to Leibniz on May 24, 1711, saying that he had not accused Leibniz of knowing the name or notation of Newton's method but that he had merely stated that Leibniz must have seen something from which he had been led to his own method. Instead of being mollified, Leibniz was outraged and declared that his honor was attacked even more openly than before; that Keill was an upstart and an unqualified judge; that he was acting without any authority from Newton; that it was the duty of the Royal Society to silence Keill; that he wanted

Newton's own opinion directly expressed; and that in Leipzig *Acts* review, "no injustice had been done to any party, as everyone had received only his due."[31] By this last comment, Leibniz made that opinion his own and at once became the agressor instead of the injured victim in the controversy.

An attempt to settle the question

A committee of the Royal Society was appointed to investigate the situation, examine the documents which had been placed in the archives, and make a decision. The committee members, appointed on March 6, 20, 27, and April 17, 1712, were Halley, Jones, De Moivre, and Machin, friends of Newton and mathematicians; Brook Taylor, a friend of Keill; Robarts, Hill, Burnet, Ashton, Arbuthnot; and Bonet, the Prussian minister. With the exception of De Moivre and Bonet, they were all Englishmen, and almost all were friends of Newton; it was scarcely an unbiased group. There was no chance for Leibniz to give his side of the story or to produce any papers he might have to substantiate his claims. Burnet wrote to Bernouilli that the committee was busy about proving that Leibniz might have seen Newton's papers.[32] The business was handled quickly, and on April 24 of the same year, the report was read. On January 8, 1713, it was published under the title, *Commercium Epistolicum D. Johannis Collinsii et aliorum de analysi promota*. The report, anonymous but probably written by Newton,[33] contains the findings of the committee and copies of the letters and papers involved in the dispute. The actual report of the committee includes a chronology of Leibniz' contacts with Newtonian influences, an assertion that Newton was the prior inventor, and a statement that Keill had not injured Leibniz by his accusation. The plagiarism issue is not touched, except to clear Newton of any possibility by declaring his priority. The report does tell of one incident in Leibniz' career

which, while it is not a direct accusation, implies that he is capable of stealing another's ideas.

February 1672/3, meeting Dr. Pell at Mr. Boyle's, he pretended to the differential method of Mouton, on being shown that it was Mouton's, insisted that it was his because he hadn't known of Mouton's doing it and had much improved it.[34]

In retrospect, the *Commercium Epistolicum* seems to have been a grossly unfair way of handling the situation. It avoided the main issue and, in effect, told Leibniz that there had been no injustice done to him because Newton had had the method before the time when Leibniz was accused of having stolen it from him; this is a meaningless and utterly illogical conclusion.

Leibniz did not see a copy of the report, but Bernouilli wrote to him about it in a letter dated June 7, 1713, adding that Newton had admitted that he didn't think of his method of fluxions until he read of Leibniz' calculus, and that Newton, when he wrote the *Principia*, had no idea of how to find the fluxions of fluxions; Bernouilli also said that Leibniz might publish his letter if he chose, as long as he did not disclose the identity of the author. Leibniz printed the letter, with comments, anonymously, without any hint as to where or by whom it was printed, under the title, *Charta Volans*. He circulated the little book among the continental mathematicians and scientists.[35] He seemed convinced that he had been injured and wrote to various mathematicians trying, unsuccessfully, to enlist them on his side; he even attacked, in a letter to the Princess of Wales, Newton's philosophy and religious orthodoxy. He wrote, but never published, *Historia et Origo Calculi Differentialis*[36] sometime between 1714 and his death in 1716, in which, writing in the third person, he gave his version of the discovery and the attempted theft of his method of the differential and integral calculus; this was apparently intended as an answer to the *Commercium Epistolicum*.

A Mr. Chamberlayne tried to make

peace between the two warring mathematicians but received the comment from Newton that Leibniz had started the fuss in 1705 with his review of the *Opticks* and a similar statement from Leibniz that it was all Newton's fault due to the *Commercium Epistolicum*. Leibniz tried one last challenge in 1716 when, through Abbe Conti, a Venetian nobleman and priest, he sent a problem to test the ability of the English mathematicians. Newton solved the difficult problem in one evening.

The quarrel gradually subsided after Leibniz' death on November 14, 1716. Bernouilli made advances to Newton, vigorously declaring that he had never said anything against the Englishman and insisting that he had not given Leibniz permission to publish any of his letters. A reconciliation seems to have followed, but there is no record of any further correspondence.

The judgment of history

It seems that the whole unsavory situation could have been avoided if both the men involved had been frank in their statements and prompt in publishing their findings. How can one account for Newton's extreme secrecy? Did he hide his methods in order to perfect them before giving them to the public? A laudable but highly imprudent motive. Did he wish to have the methods for his own exclusive use? Inexcusable selfishness. Was he trying to avoid disputes and unpleasantness? Total failure. And how can one account for Leibniz' underhanded methods? Was he striving for the honor and glory of the fatherland? Hardly, since he spent much of his time in Paris and other cities outside of Germany. Was he determined to achieve fame at any cost? But he had won renown for his work. Was he utterly devoid of honor? This is scarcely possible, for he was respected and trusted by eminent friends.

It is obvious from the confused tangle of events, the accusations and counter-accusations, the doubtful statements and

bold lies, that any definitive statement or decision is impossible. At best, one can say that Newton was probably the first inventor while Leibniz and Bernouilli were promoters and developers of the calculus. Newton seems to have invented fluxions at least ten years before Leibniz developed the calculus. There is no evidence that Newton borrowed from Leibniz; there is little evidence that Leibniz borrowed from Newton. In the hands of Newton and his followers, fluxions remained a relatively sterile theory, while Leibniz and his followers made of the calculus a powerful means of mathematical progress. To Leibniz and the Bernouillis belongs the credit for most of the vast superstructure which has been erected on the foundation laid by Newton.

Both men owed a very great deal to their immediate predecessors in the development of the new analysis, and the resulting formulations of Newton and Leibniz were most probably the result of a common anterior, rather than a reciprocal coincident influence.[37]

Claims to the invention of the calculus have been made by or for several mathematicians. Fatio de Duillier's claim was apparently ignored by his contemporaries; indeed, he seemed never to have pressed the issue himself. It has been said that credit for the calculus should go to Barrow or even to Fermat, both of whom did admirable work in laying the foundation for the later discoveries of Newton and Leibniz. But the judgment of history still stands and Sir Isaac Newton and Gottfried Wilhelm Leibniz are generally considered as the two inventors of the differential and integral calculus.[38]

A century of isolation in England

Newton's influence on English mathematics was so great that, during the entire eighteenth century, especially at Cambridge, mathematics was confined to the study of optics, gravitation, geometry, and fluxional calculus. The English savants had regarded the struggle with Leibniz and his supporters as an attempt, even a plot, of the Germans to rob Newton of the credit for his invention. In loyalty to its famous son, Cambridge chose Newton's fluxional methods in preference to Leibniz' analytical ones. For some problems, either notation may be used, but for the calculus of variations and for most modern theoretical work, Newtonian notation is impossible. In fact, even Leibnizian notation is proving inadequate for some phases of modern calculus, and that on a fairly elementary level.[39] The relative merit of the two methods was completely obscured by the quarrel over the right to credit for the invention. Personal feelings and national jealousies made the decisions, and as a result, Cambridge withdrew into a sterile isolation.

A common language is essential in the development of a science. By accepting the isolation attendant on its adoption of the Newtonian notation and refusing to make an effort to keep up with the advances made on the continent, Cambridge rendered almost sterile the efforts of the group of truly brilliant followers who had gathered about Newton. The continental mathematicians kept up with whatever English advances there were, translating the Newtonian notation into Leibnizian and thus making the work available to all. However, the English, in the isolation of injured national pride, would not do likewise as the various continental developments were published. Cambridge was out of touch with the continental mathematicians for almost a century, although the journals in which the continental findings were published were circulated widely and gratuitously.[40] As Leibniz' work was interpreted by Bernouilli, Euler, D'Alembert, Lagrange, Laplace, and others, the knowledge of the calculus spread widely among those who would listen and learn. But England was passed by; the history of English calculus led nowhere.[41]

It is not true that the differential notation was entirely unknown in England. John Craig, a friend of Newton, used dx, dy, dz and $\int$ in articles printed in 1685, 1693, 1701, 1703, 1704, 1708 and 1713. Yet in 1718, he wrote *De Calculo Fluentium* using exclusively Newtonian notations.[42]

De Moivre, in 1702 and 1703, and John Keill, in 1713, used a mixed notation of the form $\int \phi \dot{x}$, which notation was still being used in some publications as late as 1815.[43] Joseph Fenn, an Irish writer with a fine disregard to national feelings, in a *History of Mathematics*, published in Dublin sometime after 1768, used the Newtonian terms, fluxion and fluent, and the Leibnizian notation.[44]

Maclaurin in Scotland and Clairaut in France seem to be the only non-English mathematicians who used the Newtonian notation. The works of Newton and other Englishmen appearing on the continent were published as they were written, in the fluxional notation, at least until they had been "translated." It is interesting to note that there was one Dutch journal, *Maandelykse Mathematische Liefhebbery* ("Monthly Mathematical Recreations"), published in Amsterdam from 1754 to 1769, which used the Newtonian notation exclusively.[45]

The quality of the work of the English mathematicians declined rapidly after the break with the continent. Isolation accounted for part of that decline, but there was also another reason. When Newton began his work on fluxions, he was aware that he was dealing with new concepts which would be accepted only if the proofs were unimpeachable. In order to avoid having his ideas rejected because of questionable proofs, he shunned the new (1637) analytic geometry of Descartes and used only the methods of classical geometry.[46] At times he used other methods to discover theorems and derive proofs, but always he confined his final demonstrations to geometry and elementary algebra. Geometric proofs are, in themselves, adequate, but they are often labored and unnecessarily complex. Moreover, separate demonstration is required for each kind of problem; the processes are not general as they are in analysis.[47] However, long after the principles of analytic geometry and analysis were commonly accepted and freely used by the continental mathematicians, the English analysts remained true to the traditions of the master. Thus the situation stood for almost a century.

Some records of the Senate House examinations at Cambridge have come down to us, showing the general tenor of the calculus work being done at the university which was Newton's Alma Mater, a stronghold of Newtonian mathematics and physics. The 1772 examinations included:

the doctrine of fluxions, and its application to the solution of questions de maximis et minimis, to the finding of areas, . . . as unfolded and exemplified, in the fluxional treatises of Lyons, Saunderson, Simpson, Emerson, Maclaurin, and Newton . . .[48]

In 1785, the examinees were required to

find the fluent of $\dot{x}\,\sqrt{a^2-x^2}$

and to find, by the method of fluxions,

the number from which, if you take its square, there shall remain the greatest difference possible.[49]

The 1786 examination contained problems which were a little more difficult:

To find the fluxion of $x^2\,(y^n+z^n)^{1/q}$.
To find the fluxion of the mth power of the Logarithm of x.

To find the fluent of $\dfrac{ax}{a+x}$.[50]

By 1802, the students were subjected to the following types of problems:

Find the fluents of the quantities

$$\frac{d\dot{x}}{x(a^2-x^2)} \quad \text{and} \quad \frac{h\dot{y}}{y(a+y)^{3/2}}.$$

Given the fluent: $(a+cz^n)^m \times z^{pn+n-1}\,\dot{z}$, find the fluent: $(a+cz^n)^{m+1} \times z^{pn-1}\,\dot{z}$.
Required also the fluent of

$$\frac{\dot{x}\,\sqrt{a^2+x^2}}{x^3} \quad \text{and of} \quad \frac{z^{\theta}\dot{z}}{1+mz},$$

θ being a whole positive number.[51]

THE RETURN TO ANALYSIS

Towards the end of the eighteenth century, the more thoughtful mathematicians at Cambridge began to suspect the evils that were consequent upon their separation from news of continental mathematics. The logical thing to do was to adopt the Leibnizian notation and methods, but there was a sentimental objection to such an action; would not such a move be an act of disloyalty to the memory of the great Newton? Finally, in 1803, Robert Woodhouse, then a tutor and later a professor at Cambridge, wrote *Principles of Analytic Calculation*, a work which explained the differential notation and advocated its adoption. Woodhouse criticized the continental methods in some points, especially in the use of principles which were neither obvious nor enunciated. By exposing some of the errors of the system, he gave the impression that he was as much against as he was for the analytic system, thus gaining a hearing among those who would have opposed the system on the basis of those very errors. His writings seem to have been ignored by most of the professors but some of the more serious students read and wondered. As soon as they were aware of the great amount of mathematics which had been closed to them, they obtained the continental books and began to read and study. Unusual answers began to appear on some of the examinations.[52]

"A man like Woodhouse, of scrupulous honor, universally respected, a trained logician and with a caustic wit, was well fitted to introduce the new system."[53] Nevertheless, the movement might have died with him if it had not been for George Peacock, who, with Herschel and Babbage, formed an Analytical Society in 1812. As undergraduates, they habitually breakfasted together on Sunday mornings, and out of these meetings and their common interest in mathematics, the society grew.

George Peacock (1791–1858) received his B.A. from Trinity College in 1813 as second wrangler.[54] He received a fellowship in 1814 and later became a tutor.

Well loved by his students, he was a brilliant lecturer and a kindly and practical tutor, indeed a rare combination of qualities. The establishment of the University Observatory was largely due to his efforts.[55]

Sir John Frederick William Herschel (1792–1871), the son of an astronomer, entered St. John's College at Cambridge in 1809 and graduated as senior wrangler in 1813. While still an undergraduate, he wrote a paper on Cotes' theorem; he published several other papers later. He left the University about 1816 and became an astronomer and chemist.[56]

Charles Babbage (b. 1792), who entered Trinity in 1810, had had a good mathematical education before his arrival at Cambridge, having studied the works of Ditton, Maclaurin, and Simpson on fluxions, Agnesi's *Analysis* in fluxional notation, Woodhouse's *Principles of Analytic Calculation*, and Lagrange's *Théorie des Fonctions*. In 1813, Babbage transferred to Peterhouse because he wanted a chance to be first in the examinations and knew that Peacock and Herschel would surpass him if he tried to compete against them. A many-sided personality, he held a professorship, invented a machine for arithmetical processes, and wrote several scientific papers.[57]

These three eager young students, together with Maule, Ryan, Robinson, and D'Arblay, formed the original membership of the Analytical Society, whose aim was, according to Babbage, to advocate "the principles of pure d-ism as opposed to the dot-age of the university."[58] The Society published, in 1819, a translation of Lacroix' *Elementary Differential Calculus*. Peacock, who, as moderator of the Senate House examinations, was in a position to advance the cause, introduced the differential notation in the examination of 1817. In the same year, a colleague, John White, used the fluxional notation. Peacock was criticized but went on his way, feeling that the younger generation of students was ready to accept the change and that the time was right to

> reduce the many-headed monster of prejudice and make the university answer her character as the loving mother of good learning and science.[59]

The differential notation was again used by Peacock in the 1819 examinations, by Whewell in 1820, and by Peacock again in 1821. Whewell published a work on mechanics in 1819, using the differential notation; Peacock's volume on differential and integral calculus was published by the Analytical Society in 1820; Herschel's work on the calculus of finite differences, illustrative of the new method, came out in the same year; Airy, a pupil of Peacock, published *Tracts* in 1826, a work in which the new method was successfully applied to mechanics. By this time, the exclusive use of fluxions had disappeared among all but a few of the older professors.[60]

From Cambridge, the use of analytical methods spread rapidly through the rest of England. By 1830, the fluxional methods and geometric proofs had very largely disappeared, for, while the geometric demonstration is a useful auxiliary to analysis, it is almost useless as a research device.

The results of the Newton-Leibniz controversy in terms of the personal pain and mental disturbance suffered by the two principal protagonists cannot, of course, be adequately judged. The effect of the controversy on the mathematical world seems to be twofold. As far as credit for the discoveries is concerned, today the two men are honored equally as two independent inventors. Concerning the development of analysis, England seems to have been the loser. The world of mathematics progressed. German and French mathematicians established reputations for themselves and their countries, while England remained insular and isolated. What was the cost to English mathematics? Perhaps no one will ever know. British mathematicians and scientists have done much in the last hundred years for the honor of England and the advancement of knowledge. Nevertheless, one cannot but regret the irreparable loss occasioned by that "dark age" in intellectual history.

Footnotes

[1] Carl B. Boyer, *The Concepts of the Calculus* (Wakefield, Mass.: 1949), p. 187.

[2] *Ibid.*, pp. 187–88.

[3] Sir Isaac Newton, *The Method of Fluxions* (London: 1736), p. 19.

[4] *Ibid.*, p. 21.

[5] *Ibid.*, p. 19.

[6] *Ibid.*, p. 25.

[7] *Ibid.*, p. 20.

[8] J. Edleston (ed.), *Correspondence of Sir Isaac Newton and Professor Cotes* (London: 1850), p. 169.

[9] "Commercium Epistolicum," *Philosophical Transactions*, No. 342, January, February 1714/15 (London: 1717), p. 204.

[10] Felix Klein, *Elementary Mathematics from an Advanced Standpoint*, Part I (New York: 1932), p. 215.

[11] Florian Cajori, *A History of Mathematical Notation*, Vol. II (Chicago: Open Court Publishing Co., 1929), p. 181.

[12] Sir Isaac Newton, *Mathematical Principles of Natural Philosophy* (Berkeley, Calif.: University of California, 1947), pp. 655–56.

[13] John Theodore Merz, *Leibniz* (New York: 1948), p. 60.

[14] David Brewster, *The Life of Sir Isaac Newton* (New York: 1831), pp. 175–76.

[15] Augustus De Morgan, *Essays on the Life and Work of Newton* (Chicago: Open Court Publishing Co., 1914), pp. 95–96.

[16] Arthur S. Hathaway, "The Discovery of the Calculus," *Science*, New Series, Vol. L, No. 1280 (July–December, 1919), pp. 41–43.

[17] Merz, *op. cit.*, pp. 54–55.

[18] Hathaway, *op. cit.*, p. 42.

[19] Hathaway says that Leibniz' methods here described are like the methods of German propaganda in World War I and that Leibniz deserves no honor at all even if he were an independent discoverer because "he does not come into court with clean hands." *Op. cit.*, p. 43.

[20] De Morgan, *op. cit.*, p. 25 (note).

[21] Merz, *op. cit.*, pp. 58–59.

[22] Klein, *loc. cit.*

[23] Newton, *Mathematical Principles, loc. cit.*

[24] *Ibid.*

[25] J. W. N. Sullivan, *Isaac Newton* (New York: 1938), p. 229.

[26] Gottfried Wilhelm Leibniz, *The Early Mathematical Manuscripts of Leibniz*, J. M. Child (ed.) (Chicago: Open Court Publishing Co., 1920), p. 8.

[27] Sullivan, *op. cit.*, p. 232.

[28] *Ibid.*

[29] See below.

[30] Sullivan, *op. cit.*, p. 234.

[31] Brewster, *op. cit.*, p. 189.

[32] De Morgan, *op. cit.*, pp. 27–28 (note).

[33] Augustus De Morgan, "On the Authorship of the Account of the *Commercium Epistolicum* in the *Philosophical Transactions*," in *Philosophical Magazine*, 4th series, Vol. 3 (1852), pp. 440–43.

[34] "Commercium Epistolicum," *Philosophical Transactions*, p. 183.

[35] Brewster, *op. cit.*, pp. 192–93.

[36] Leibniz, *op. cit.*, pp. 22–57.

[37] Boyer, *op. cit.*, p. 188.

[38] For a discussion of the validity of other claims, see Florian Cajori, "Who Was the First Inventor of Calculus?" *American Mathematical Monthly*, XXVI (1919), 15–20; and J. M. Child, "Barrow, Newton, and Leibniz, in Their Relation to the Discovery of the Calculus," *Science Progress*, XXV (1930–31), 295–307

[39] R Creighton Buck, *Advanced Calculus* (New York: McGraw-Hill Book Co., Inc., 1956), pp. 58–59.

[40] W. W. Rouse Ball, *A History of the Study of Mathematics at Cambridge* (Cambridge: 1889), p. 98.

[41] *Ibid.*

[42] Florian Cajori, "The Spread of Newtonian and Leibnizian Notations of the Calculus," *Bulletin of the American Mathematical Society*, XXVII (June–July, 1921), 453.

[43] *Ibid.*, p. 454.

[44] *Ibid.*

[45] *Ibid.*, p. 455.

[46] Ball, *op. cit.*, p. 69.

[47] *Ibid.*, p. 98.

[48] *Ibid.*, p. 192.

[49] *Ibid.*, pp. 195–96.

[50] *Ibid.*

[51] *Ibid.*, pp. 200–09.

[52] *Ibid.*, p. 119.

[53] *Ibid.*

[54] Wrangler—an honors student, in the first class in the mathematical Tripos. First ranking man is designated as senior wrangler, next as second, third, fourth, etc., wranglers.

[55] Ball, *op. cit.*, p. 124.

[56] *Ibid.*, pp. 126–27.

[57] *Ibid.*, pp. 125–26.

[58] *Ibid.*, p. 126.

[59] *Ibid.*, p. 121.

[60] *Ibid.*, p. 122.

Bibliography: *Historical Overview*

Graves, G. H. Development of the Fundamental Ideas of the Differential Calculus. 3 (December 1910): 82–89.
   A discussion of various approaches for introducing the fundamental concepts of calculus.

Sanford, Vera. Sir Isaac Newton, 1642–1727. 26 (February 1933): 106–9.

______. Gottfried Wilhelm Leibniz, 1646–1716. 26 (March 1933): 183–85.

______. Leonard Euler, 1707–1783. 27 (April 1934): 205–7.

Wren, F. Lynwood. Curiosity and Culture. 50 (May 1957): 361–71.
   Cites contributions of such men as Copernicus, Descartes, Galileo, Kepler, Leibniz, Newton, and Pythagoras to the understanding of the physical world.

Jones, Phillip S. Sir Isaac Newton: 1642–1727. 51 (February 1958): 124–27.

Shelton, Julia B. A History-of-Mathematics Chart. 52 (November 1959): 563–67.

Eves, Howard. The Bernoulli Family. 59 (March 1966): 276–78.

Greitzer, Samuel L. Credit Where Credit Is Due! 60 (February 1967): 155–57.
   Specific recommendations for the use of history in teaching mathematics; contains bibliographical references.

Mulcrone, Thomas F., S.J. A Plea for the Terminology "Flexpoint." 61 (May 1968): 475–78.
   Remarks concerning a proposal to adopt the term *flexpoint* for inflection point. Discussion is based on Latin derivations and historical usage.

Read, Cecil B. Anomalous Mathematical Nomenclature. 62 (February 1969): 121–25.
   Remarks concerning erroneous credits for the development of mathematical ideas.

Lick, Dale W. The Remarkable Bernoulli Family. 62 (May 1969): 401–9.

2

Pedagogical Overview

The concern of mathematics educators in general and calculus instructors in particular is two-pronged—the development of skill proficiency on the one hand and of concept understanding on the other. Some tried-and-tested approaches to calculus instruction are described in this section.

Parker discusses general teaching objectives for a calculus course and gives concrete suggestions for their realization. Cummins describes an experimental, college-level calculus course that was taught using a student experience-discovery approach. The author feels that students taught by the methods outlined here not only develop proficiency in manipulative skills but also experience the joy of mathematical discovery. Finally, Flanders provides detailed analyses of specific calculus problems. Attention is focused on the background that students need for success in calculus. These analyses can be used effectively in actual classroom teaching, since students' learning is reinforced by continued recall and reiteration of material already learned.

The contents of this section are applicable to the calculus course as a whole. Additional concept-oriented, pedagogical suggestions are found throughout this book.

The Teaching Objectives in a First Course in the Calculus

By JAMES E. PARKER

Knoxville College, Knoxville, Tennessee

THERE IS a great deal being said nowadays about course objectives. It is said that the objectives for any course in the curriculum should be clearly defined and that teacher and student alike should be conscious of these objectives at all times. This concern has lead to a re-examination of the objectives for courses taught by the writer. This paper is particularly concerned with the objectives for a first course in differential and integral calculus. Stated briefly, the question at hand is this *what are my objectives in teaching a first course in the calculus?* Three major objectives are found to be inherent in this case. These are: (1) to give the student an understanding of the fundamental concepts of the calculus and a point of view relative to the historical background out of which these concepts grew; (2) to develop proficiency in the manipulative skills of differentiation and integration; and (3) to develop abilities in making practical applications of the principles learned. A few words will be said about each of these objectives relative to the methods used in attempting to realize them. It is hoped that the following statements will arouse some interest and will bring forth some comments as well as criticisms from individuals more experienced than the writer. In this way the writer, and possibly others, may be greatly helped in setting up goals toward which to work.

In the first place, we are assuming that a knowledge of the fundamental concepts of the calculus is essential to a successful effort to the study of the calculus. The concepts of a function, a variable, increment, limit, and continuity are absolutely necessary stepping-stones for the beginning student, and the well-equipped teacher "is prepared to put into proper perspective before his students the most fruitful concepts of his field."[1] These are the words of Dr. C. E. Van Horn, who says further:

In introducing students to new concepts, nothing can be safely taken for granted. The instructor must know that the students are clear on all the little points that surround the subject.

An instructor must take time to encourage habits of clear thinking. And nothing, perhaps, is more conducive to the cultivation of such habits than to be clear-cut in characterizing logically the concepts with which one may be dealing.

It is indeed difficult to imagine a student making strides of progress if he is lacking in a clear knowledge of such a concept as the limit, for example. It seems that first and foremost in the minds of calculus teachers must be a definite aim to teach the underlying concepts of the calculus. This should not be done as a separate unit of work but the concepts should be introduced as the need for each arises.

One method of realizing this objective is to have the student write brief papers in which he is required to discuss certain important concepts which are being studied at the time. The student should be encouraged to make crystal clear the concepts discussed. Through complete statements and through a strict choice of words the student soon develops the ability to give real meaning to these concepts. Such a paper enables the teacher to locate discrepancies in the student's understanding, and, when properly handled by the teacher, reveals to the student his own errors in thought, thus allowing for self-evaluation in growth.

Secondly, we must always consider the

[1] Van Horn, *A Preface to Mathematics*, Chapman and Grimes, Inc., 1938.

skills accompanying the study of practically every elementary course in mathematics. A successful calculus student is certainly not remote from these skills. Calculus can hardly be taught without some consideration of a degree of proficiency in the skills of differentiation and of integration. We should, however, take care that we do not over-emphasize this aspect of teaching. Too often courses in elementary mathematics turn out to be no more than a routine process of *stereotyped problem solving*. Apparently the better way to approach this phase of the work is to work on a selective basis. That is to say, select problems which best suit the needs of the hour and best illustrate the principles at hand. This is a more flexible method than that of merely following the text through problem by problem. This method may also sidetrack any tendency for teacher and student to become textbook slaves. There is, on the other hand, a danger of under-emphasis of the development of the basic skills. An over-zealous teacher is likely to become bored by what may seem to him drill work. Most calculus textbooks have a large number of problems. This teacher is likely to feel that too much emphasis is placed thereon. He should realize that a student can hardly find his calculus at his command without proficiency in these basic skills. The ability to make practical applications of calculus is in a very large measure dependent upon these skills. For after all, practical application of any phase of the calculus is a process of finding a satisfactory answer to some problem. This leads to our third objective: namely, to develop abilities to make practical applications of the principles learned.

For the past quarter of a century or more we have witnessed a decline in the teaching of mathematics in our secondary schools. Educators have been prone to believe that little mathematics is needed by the average man. Consequently, what has actually happened is that the selective few whom they consider in need of the subject have suffered in their early work. This

is particularly true of the small schools. If on the one hand, mathematicians have been prone to over-emphasize pure mathematics, educators have, on the other hand, failed to recognize the value of pure mathematics in its service capacity, its advisory capacity. We are here considering the phrase *practical applications* in a much broader sense than the mere use of the subject as a tool for engineers. If we examine the records we will doubtless find that the work of mathematicians has very often been a reservoir for many hints or helps along practical lines. For a thorough discussion of this aspect of the teaching I refer you to Mr. Thornton C. Fry's article "Industrial Mathematics," *The American Mathematical Monthly*, vol. 48, June–July 1941, Part II, Supplement. Here are listed some fine examples of the very thing of which we are speaking. These are supported by the best of evidence. Indeed all mathematics is practical in several ways. To be sure the word practical infers usefulness. But a thing may be useful in a number of ways.

Teachers of calculus should be alert in attempting to find ways of bringing home to the student the various applications of his subject. This can be done in seemingly insignificant ways. For example, in working the following problem in maxima and minima a student of farm experience found that in a small way his calculus could help him out on the farm:

An open trough is to be made from a long rectangular shaped piece of metal by bending up the long edges so as to give the trough a rectangular cross section. If the width of the piece is 1 foot, how deep should the trough be made in order that its carrying capacity may be a maximum?[2]

This may seem rather trivial, but the point is this. This young man saw the usefulness of his calculus and subsequently was motivated to study calculus more conscientiously. Students as far along as this may

[2] Ford, W. B., *A First Course in the Differential and Integral Calculus*, Revised edition, Henry Holt and Company, New York, 1937.

often need to be motivated in this way. As a matter of fact, college students are more and more demanding insights into the usefulness of their subjects, and the calculus teacher has an unique opportunity in this connection.

The writer is, of course, teaching in a liberal arts college, but he feels that any student taking the course prescribed for above should find himself ready and capable in whatsoever field of endeavor he may find himself. The statements made herein are merely suggestive. No attempt has been made to write an extensive discourse on the topic. The purpose is to bring to a focus the objectives set forth by the writer.

A student experience-discovery approach to the teaching of calculus

KENNETH CUMMINS, *Kent State University, Kent, Ohio.*

Students of calculus experienced the thrill of discovery and possessed superior knowledge of the theory of calculus when taught by the methods discussed in this article.

INTRODUCTION

IN THE LAST several decades there has arisen much concern over the importance of understanding and meaning in mathematics as developed by the student himself. One need only refer to such studies as those of Fawcett,[1] Sobel,[2] Schaaf,[3] and others to note experimentation at the high school level on encouraging a maximum of student discovery and student organization of materials. Results of these investigations suggest that the student gains deeper understanding when he is permitted to organize mathematical ideas in his own way and that this is accomplished with no loss of problem-solving proficiency. Too, and perhaps most important of all, the student who seeks to defend *what he himself has formulated* experiences the deep-seated joy of creativity and develops a realistic insight into how mathematics grows.

ORIGIN OF THE STUDY

In attempting to encourage advanced mathematical study by interested and able students the writer devised a one-semester course in polynomial calculus as a part of a four-year program in secondary mathematics.[4] Succeeding years brought forth improved study guides until it became possible, it seemed, for the students to see calculus in its unveiled beauty and clarity and with none of the mystery that so often shrouds the heart of the subject and causes manipulations to be performed in a meaningless way.

Indeed, it has been found that this captivating problem of *teaching* the calculus has been of deep-felt interest for some time. As early as 1907 Saxelby, in *A Course in Practical Mathematics*, recognized the problem of helping students *understand* the calculus, and in writing of his approach to differentiation suggested that "it too often happens that a student . . . acquires a merely fatal facility in differentiation, regarding it as a mechanical juggling with symbols but having no conception of its relation to experience."[5] Saxelby further urges the teacher to consider the immense value of experimental or numerical or arithmetical or graphical methods and then adds that

> . . . this intuitional direct vision method is intended, not to take the place of, but to prepare the way for, a more rigorous analytical study of the subject. . . . The most natural method of advance is by a series of successive approximations to logical rigor, and, in fact, this is the way in which the subject has actually grown up.

---

[1] Harold P. Fawcett, *The Nature of Proof*, Thirteenth Yearbook (Washington, D. C.: National Council of Teachers of Mathematics, 1938).

[2] Max A. Sobel, "Concept Learning in Mathematics," THE MATHEMATICS TEACHER, XLIX (October 1956), 425–30.

[3] Oscar Schaaf, "Student Discovery of Algebraic Principles as a Means of Developing Ability to Generalize" (Doctoral dissertation, Ohio State University, 1954).

[4] New Washington (Ohio) High School.

[5] F. M. Saxelby, *A Course in Practical Mathematics* (London: Longmans, Green, and Company, 1907), p. v.

... The process by which the science itself was formed is also the most natural for the mind of the student.[6]

Pólya urges that we examine a second, often ignored, facet of mathematics—"mathematics *in statu nascendi*."[7] He says that "mathematics in the process of being invented has never before been presented in quite this manner to the students, or to the teacher himself, or to the general public."[8]

Teacher-scholars from both the past and the present have sounded a call for a challenging experiment. Is it really possible to conduct a planned course in beginning calculus at the faster-paced university level and to utilize as much as possible discoveries of students themselves? Will such an approach increase understanding? The remainder of this article is largely a story of an attempt to test the hypothesis that a student experience-discovery approach to calculus in the university yields results every bit as promising as those under similar conditions on the secondary level.

The study planned

Accordingly an experiment was planned which would put this hypothesis to test. First, it was necessary to arrange a sequence of topics for a beginning-quarter course in such a way that the contributions of students could be advantageously utilized. Materials were arranged to develop understanding in the use of some of the fundamental ideas before these concepts were subjected to critical discussion by the class or before results suggested as hypotheses were finally deduced. One section of the course was devoted to developing the calculus as a deductive system.

Secondly, there was constructed a series of study-guide sheets containing questions, suggestions, and exercises by which the student, either independently or with the help of class discussion, could arrive largely by his own efforts at some methods

and facts of the calculus.[9] It was planned that the text for the course, Hollis R. Cooley's *First Course in Calculus*,[10] would be used for problems other than those of the study guides, as a mathematics source book, and/or as a means for the student to compare "his own book" with that of another author.

It was also desirable to arrange for experimental and control sections in the first-quarter calculus. The one-quarter experimental sections were conducted for two quarters in an atmosphere rich in encouraging discovery, whereas the control groups were taught more or less traditionally by capable men of long university teaching experience. The same text was used in all sections.

To provide some means for comparison of the two experimental and the two traditional sections and to prepare for a statistical study of results, the investigator used a calculus pretest constructed by the writer, the American Council of Education (ACE) Psychological Test, and the average grade in previous mathematics courses.[11] For brevity, tables of comparison are not included in this article.

At the close of the quarter each section took the two-hour examination made by the experimenter for his section and the examination made by the other section's instructor for his section. Further evaluation of the experimental course was attempted by the use of written student reactions and by noting student thought in a short paper on "What I Conceive the Calculus to Be."

At the beginning of the course, there was a short discussion of situations and problems which involve rates of change. Quite immediately the class was con-

---

[6] Saxelby, *op. cit.* pp. v, vi.

[7] G. Pólya, *How to Solve It* (Princeton: Princeton University Press, 1946), p. vii.

[8] *Ibid.*

[9] During the writer's talk at the 18th Christmas Meeting of the National Council of Teachers of Mathematics, sample copies of some of these study guides were distributed. Other requests were filled by mail.

[10] Hollis R. Cooley, *First Course in Calculus* (New York: John Wiley and Sons, Inc., 1951).

[11] A few other comparisons were made involving, among other things, the time gap between the study of analytical geometry and calculus, age, military experience, and attitude toward mathematics.

fronted with computing the speed of the body labeled $(m+f)$ in the apparatus shown in the figure as it fell under the action of a force f. The speed at the end of the second second ($t=2$) was chosen for consideration. Data were taken as members of the class volunteered to mark distances fallen at the end of each second as determined by a swinging pendulum. The group suggested that quotients of smaller and smaller distances from the point at which $t=2$, divided by corresponding time intervals such as $\frac{1}{2}$ sec. or $\frac{1}{4}$ sec., would give better approximations to the speed at $t=2$. The difficulty of marking distances at times indicated by pendulums swinging with increased rapidity, however, soon provided convincing evidence that this method was not very feasible.

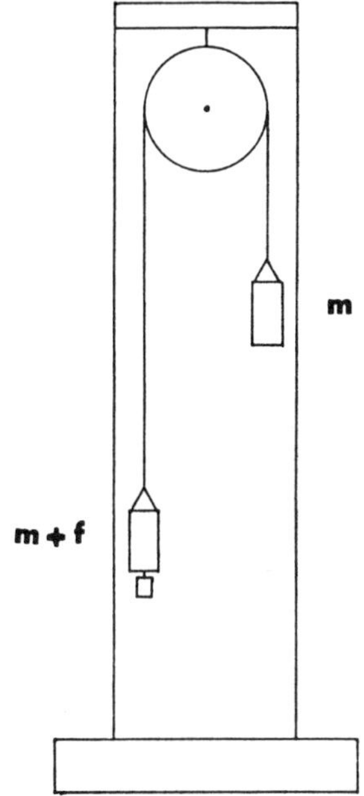

This difficulty led to the acceptance of *idealized expressions* to describe nonuniform motion such as the well-known relation $s=16t^2$. The students suggested the same method of attack, only this time average speeds were computed by the use of a *formula* to arrive at a sequence of approximations to the speed at $t=2$ sec. In fact, using times $t=2$, $t=2\frac{1}{2}$, $t=2\frac{1}{4}$, $t=2\frac{1}{8}$ secs. $\cdots$ to compute distances through

which the body fell at the end of these times, and from these distances to arrive at average speeds during various time intervals, one obtains a sequence of average speeds 72, 68, 66, 65, $64\frac{1}{2}$, $\cdots$ feet per second. The answer to the problem suggested by this sequence is 64 feet per second. Here it was emphasized that we did not know (a) if this sequence of average speeds really converged, (b) what the "limit" really was, or (c) if this "limit" was the answer to our problem. This uncertainty nevertheless gave rise to some hypotheses and definitions, and a logical structure was begun, although it was felt that changes might come upon re-examination. Numerical exercises were suggested that investigated in an introductory way other manners of making the time interval smaller.[12] Some quite ingenious methods were suggested and worked out by the students. Student imagination, student discovery, and student contributions frequently reached a pleasing peak.

To lessen the amount of numerical computation which had served well to develop fundamental ideas, the average speed from $t=2$ to $t=2+h$ was used. Then the speed at $t=2$ was taken to be the

$$\lim_{h \to 0} \frac{16(2+h)^2 - 16(2)^2}{h},$$

and the students gained practice in a first step in abstraction.

From student suggestions in the various groups actually came the idea that one might compute the speed at *any* time $t=t_1$ and hence work "many problems in one" by this procedure. Speeds for various times and for various expressions of distance could then be found by replacing t_1 by some numerical value for time.

Of course this speed problem was carefully examined, and at the same time the problem of computing the slope of a curve

<hr>

[12] In one case, $t=2$ was approached through the times $t=3$, $t=2\frac{1}{2}$, $t=2\frac{1}{4}$, $t=2\frac{1}{8}$, $\cdots$; in another case, the sequence was $t=3$, $t=2.1$, $t=2.01$, $\cdots$: in still another, $t=1$, $t=1\frac{1}{2}$, $t=1\frac{3}{4}$, $t=1\frac{7}{8}$, $\cdots$.

was considered. It did not take long for the students to recognize elements common to the two problems, and from this consideration emerged the general "rate-of-change" idea. The result of computing this rate of change of y with respect to x was indicated by R_xy.

The discovery of ways to compute R_xy at sight, various uses of R_xy, the properties of R_xy at maximum and minimum points on a curve, and the use of R_xy to approximate changes in y relative to a given change in x at $x=x_1$ were all developed by inductive exercises and discussion on the guide sheets.

Not until the more critical study of R_xy and its definition as the limit of a sequence of quotients,

$$\frac{f(x_1+h)-f(x_1)}{h},$$

as $h\rightarrow 0$, was the symbol changed to that of D_xy.

In like manner, the idea of the *integral* had a humble numerical beginning. An intuitive notion of area under a curve was accepted, and attempts were made to find the area bounded by a curve such as one whose equation is $y=x^2$, $x=1$, $x=3$, and the x-axis. Some students suggested the use of sums of areas of rectangles to make approximations, but it seemed that students who attempted to use the sums of areas of *trapezoids* fared better in ease of setting up numerical sequences to lead to a number which they would present as the area. Results of the use of trapezoids are shown in Table 1. As a class exercise, it

TABLE 1

SUMS OF AREAS OF TRAPEZOIDS IN APPROXIMATING (INTUITIVE) AREA UNDER $y=x^2$ FROM $x=1$ TO $x=3$

Number of trapezoids used	Sum of areas of these trapezoids
1	10
2	9
4	8 3/4
8	8 11/16
16	8 43/64

was noted that the sum of areas for any number of trapezoids could be expressed as

$$S=10-(1+1/4+1/16+1/64 \cdots)$$

and that this could be evaluated by the aid of the well-known formula for the sum of an infinite geometric progression, $S_\infty=a/(1-r)$, whence $S=8\frac{2}{3}$ square units. Although the use of trapezoids was preferred in computational work, area was *defined* as the limit of a sequence of sums of areas of rectangles as the number of rectangles increased and as the widths of these rectangles decreased. In a similar way were considered problems of "work done" when the force is variable.

In the critical study of these concepts students noted that area and work as defined are both limits of sequences of sums of *products* (analogous to *quotients*, which lead to the derivative), and from these two ideas a definition was abstracted for a concept for which, as yet, we had no name! For some time this was called the "area-computing" process, and a symbol such as $A(x^2)\big|_0^2$ was used to signify area (and also work).

As the course progressed, the usual theorems on sequences, limits of functions, and derivatives were deduced. The limit of a sequence of sums of products was given the name "integral," and theorems on integrals were formulated. From the very definition of the integral, the values of

$$\int_a^b kdx, \quad \int_a^b kxdx, \quad \text{and} \quad \int_a^b kx^2dx$$

were worked out and listed as theorems. It was then *hypothesized* that

$$\int_a^b kx^ndx$$

had the value

$$\frac{k(b^{n+1}-a^{n+1})}{n+1}.$$

Striking relations between these results and the antiderivative of kx^n were discovered and noted by the students throughout all this study, and a feeling developed that some great fundamental link between the derivative and the integral was in the making. This sense of "something about to happen" came to a climax with the fundamental theorem of the calculus

$$D_t \int_a^t f(x)dx = f(t),$$

for $f(x)$ continuous in $a \leq x \leq t$, and with the corollary that

$$\int_a^b f(x) = F(b) - F(a)$$

where $D_x F(x) = f(x)$.

It should be mentioned that topics were so timed in the planning of the work that while some ideas were being refined through discussion and investigation other new thoughts were being introduced at a more elementary level; hence several different topics were under consideration at the same time. This was done to encourage careful and slow nurturing of ideas.

Student growth and contributions throughout the unfolding of the calculus created an inspiring morale in the group as the classroom became a center for discussion, practice, drill, guidance, and investigation. Several of the definitions, hypotheses, and theorems were designated by the names of students who first suggested or deduced them, to help to emphasize how mathematics grows.

RESULTING CONTENT

Although a first-quarter course in the calculus had been previously outlined, some minor changes resulted while the work was conducted under this student experience-discovery approach. As one would expect, there were really two facets in content: (*a*) the sequence of topics studied, and (*b*) some realizations of more intangible nature—an awareness of logical structure in the calculus and the role of inductive and deductive approaches to mathematics. The resulting topical content is shown in Table 2.

TABLE 2

TABLE OF CONTENTS OF EXPERIMENTAL COURSE

I. *Problems of the Calculus*
   A. The General Rate Problem
      1. Estimating experimentally the speed of a falling body. Exercises.
      2. Estimating the speed when a formula for distance is known. Exercises. Fundamental terms. Exercises.
      3. Estimating speed with less numerical computation. Functions. Limits.
      4. Estimating speed at any time t_b. Exercises.
      5. Finding the slope of a curve. Graphs. Maxima and minima. Exercises.
      6. General rate of change. Functions. General picture. Exercises.
      7. Uses of changing rates. Related rates. Errors. Small changes.
      8. General review exercises.

   B. The General Area Problem
      1. Approximating area (intuitive) under a curve. Numerical methods. Other methods. Exercises.
      2. More work on areas. Exercises.
      3. Approximation of work done on a spring. Exercises.
      4. Use of approximating *functions*. Exercises.

   C. The Fundamental Problems Recapitulated

II. *A Systematic Look at the Calculus*
   A. Sequences. Limits. Theorems.
   B. Limits of functions.
   C. Continuous functions.
   D. Review.
   E. General review exercises.

III. *The General Rate Problem Treated*
   A. The Derivative
      1. Definition developed from a previous process. Some theorems. Exercises.
      2. Making differentiation easier. Differentiation of products, quotients.
      3. Differentiation of a function of a function. Exercises.
      4. Inverse functions. Parametric functions.
      5. Implicit functions.
      6. Operation inverse to differentiation.
      7. General review exercises.

   B. Second and Higher Derivatives
      1. Derivatives of derivatives and their interpretation. Use.
      2. Derivatives and the study of rectilinear motion. Exercises.
      3. Review exercises.
      4. Higher derivatives of parametric functions.
      5. Higher derivatives of implicit functions.

C. Inverse Rate Problem Continued
  1. How can inverse differentiation be used?
  2. Given slope to find equation of a curve. Finding the length of a curved segment.
  3. A general formula for antidifferentiation of ax^n.

IV. *The General Area Problem Treated*
  A. The Definite Integral
    1. Defined as a name for something we have developed before. Theorems. Refining our thinking. Fundamental terms.
    2. Fundamentals, theorems, applications.

  B. Evaluation of Integrals
  The mean-value theorem. Applications of integrals. The fundamental theorem of the calculus.

  C. Some Applications of Integrals
  Areas. Work. Volumes. Force of liquids on a vertical wall.

V. *General Review*

Throughout the course there was much emphasis on delineating undefined terms, defined terms, hypotheses or postulates, and deduced statements or theorems. In fact, in every experimental section some fundamental philosophical questions often arose, such as: "Is calculus based upon things 'approaching'—if so, isn't this a rather weak foundation?" Here was an opportune time to discuss again the structure of mathematics. One student commented aptly that "Our faith that a certain number is the answer to a speed or to an area problem is just as firm as our faith that a certain sequence approaches a certain limit." On another occasion there arose the question of distributivity of the "area-computing process"—is $A(x^2 + x^3) = A(x^2) + A(x^3)$? The study of this question was another rewarding experience in the group's mathematical adventure and development.

Each student kept a notebook that he organized in his own way. A central core of hypotheses, definitions, and theorems, however, was quite common to all such efforts.

The *skills* developed in traditional first-quarter calculus were nurtured in this course also. In this approach, however, the student gained more practice in *numerical work* as he was being guided to abstractions.

EVALUATION

As indicated before, several kinds of evaluation were attempted. The cross-test made for the experimental sections and taken by both the experimental and traditional groups consisted of short-answer items and discussion questions that were designed to measure understanding. It also contained problems. The test made for the traditional sections by their teachers, and taken by both groups, largely contained problems. Each teacher used these tests in whatever manner he wished for his own course. The tests were mixed into a packet and graded uniformly by the writer. The results of this grading are the scores which appear in this study. Results from the winter quarter cross-testing program are found in Table 3.

TABLE 3

ARITHMETIC MEANS OF SCORES ON TESTS (WINTER QUARTER)

Section taking test	Score on Test One made for Experimental Section One. Total score possible 272	Score on Test A made for Traditional Section A. Total score possible 180
Section One (experimental)	217	130
Section A (traditional)	131	130

It seemed advisable to give further study to these final test results in the light of initial data on previous grades, the pretest, and the ACE scores, in an attempt to make some conclusions on performance independent of the initial data. Accordingly, data of the traditional and experimental groups were combined to develop regression equations which would predict for each student an expected score on Test One and on Test A on the basis of the three preliminary test scores of both sections. In fact, it turned out that the ACE and calculus pretest scores, at least as used here, had no significant effect on the Test One and Test A scores. The results showed that:

   a. The students in the experimental group scored on the average 27.10 points higher on Test One than would be expected on the basis of their preliminary test scores (significant at the 1% level).

   b. The students in the traditional group scored 51.59 points lower on Test One than would be expected (significant at the 1% level).

   c. The differences were not significant on Test A.

The results indicate that the method of teaching under examination was especially effective in promoting a deeper understanding of the calculus and that this gain was not at the sacrifice of proficiency in manipulations and applications. It is possible that these differences in achievement were due to factors unknown to the experimenter, but the statistical analysis indicates that they were not due to student variations in previous grades, mathematical preparation as measured by the calculus pretest, or innate ability as indicated by the ACE scores.

Since the results from the winter quarter were so positive, a statistical study was not made on the test scores at the close of the spring quarter. The data "in the raw" indicated similar results, however.

In addition to statistical studies on cross-tests of proficiency and understanding, there was an attempt to evaluate the effectiveness of this approach to the calculus by searching out student reaction on certain items. Some of these items are listed in Table 4.

Inquiries were also made on student awareness of logical relation in the calculus, on experiencing how mathematics grows, on sensing the power of the calculus method, on the enjoyment of mathematics, and on whether, and at what points,

TABLE 4

NUMBER OF SATISFACTORY RESPONSES TO SOME TEST ITEMS ON LOGICAL RELATIONS

Test item	Nature of expected response	Number of satisfactory responses from	
		Experimental Sections	Traditional Sections
$D_x y = \lim_{h \to 0} \dfrac{f(x_1 + h) - f(x_1)}{h}$ What is the limitation placed upon h?	$h \neq 0$	36	10
The relation $D_x(f+g)$ comes from what basic theorem on limits?	$\lim_{n \to \infty} (a_n + b_n) = A + B$	24	7
What would you suggest as the concept most basic to the calculus?	Sequences or limits	24	5
Deduce the quotient rule, the product rule, or the chain rule.	(A correct deduction)	25	1
Explain briefly or describe the logical thread which connects the process of differentiation with that of integration.	Antidifferentiation and the fundamental theorem	18	0
Total number of responses.		38	24

the student experienced a feeling and thrill of discovery. Without attempting to classify them in this article, sundry quotations taken from student responses will help communicate student reaction:

1. "They (the derivative and the integral) both have the same beginning in the theories of limits. They are both limits, only the derivative is the limit of a sequence of quotients and the integral is the limit of a sequence of sums of products."
2. "We start with sequences and go from there to the [idea of] rate of change, and then to the derivative and the antiderivative. On the other side of the picture we go from sequences up through areas to the integral. Here we go through the mean-value theorem for integrals to the fundamental theorem. The fundamental theorem, therefore, is the link which holds the two chains together."
3. "I could see the calculus grow from the beginning arithmetic methods to the more involved theorems."
4. "Our ideas started on a 'man-on-the-street level' and then grew."
5. "The completeness of our calculus structure impresses me. It is to me like building a house—afterwards I don't just look at the roof, but at all the beautiful parts of the house under the roof."

Perhaps the most important measure of all is the student evaluation of this student experience-discovery approach. A summary of responses is given in Table 5.

TABLE 5

A SUMMARY OF FIFTY-THREE RESPONSES TO "DO YOU THINK THIS APPROACH IS REALLY DIFFERENT?"

Student Responses	Frequency
I understand *reasons* for doing, rather than doing these mechanically	22
New work is developed from the student from his own knowledge	9
Students helped to develop the course; were not merely passive in acceptance	5
Class more "human"; better student-teacher relation	4
We saw how things develop and we co-operated to bring out a new idea	4
We saw a problem and then saw progressively simpler methods to find the answer	4
This approach gives one a good beginning	3
My interest has been aroused more than ever before	2

Some statements seem to convey more than it is possible to suggest when they are condensed in a summary table:

1. "[This approach is different] because we really got to the very basic beginning which is something I have really not done before; most other courses are such that we work a lot of problems with very little theory."
2. "It stresses the reasons and theory behind the problems. This is something *I* have never seen and have always wanted in a mathematics course."
3. "[This approach is different] mainly because it is the first time in all my mathematics that the very basic material was taught so for once I know what I was really studying."
4. "It is different in many ways. The most significant is the fact that the student is helped to discover the principles for himself, thus giving a clearer understanding."
5. ". . . this makes [the calculus] a part of him [the student] and it stays alive for him . . . this is the best understood course, the finest, and the most clear course of any kind that I have ever taken."
6. "It used to be that I did not know how they got what they got in mathematics, but now I do."
7. "This approach makes the mathematics 'alive.' It is living and growing! Some students I talk to just differentiate—they do not know what they are doing, nor do they understand what a derivative is."
8. "This method seems to take the magic or mystery out of mathematics."

SUMMARY

It is generally agreed that concern for understanding rather than manipulative skill alone is a matter of utmost importance in encouraging the attainment of the quality which a mathematics education should provide.

Hence there is proposed in this study a student experience-discovery method whose methodology has characteristics as follows:

1. General problems find their initial setting in natural situations.
2. The students attack these problems with the means they know—often arithmetic.
3. Results are studied, examined, discussed; definitions and hypotheses are made.
4. Abstract considerations flow naturally out of the concrete.
5. The teacher guides the student to the consideration of more profound questions.
6. Statements accepted as hypotheses remain so until deduced as theorems with what is being *assumed* clearly delineated and set apart from what is being deduced.

7. Student initiative, discovery, and suggestions are encouraged and utilized to help develop the calculus.

A test of this method in the classroom results in some tentative "good news" for the student. On the basis of this experiment there can be proposed the working hypothesis that students taught by this "discovery" approach have these advantages:

1. They do as well on problems and manipulative skills as those with traditional instruction but, in addition, they have increased understanding.
2. They possess a superior knowledge of the fundamental theory and logical relations among parts of the calculus.
3. They experience the thrill of discovery and the satisfaction of producing results through creative effort—all of which lead to greater enjoyment of mathematics and a deeper understanding of its nature and use.
4. They express ideas of the calculus in their own language and they undergo the stimulating and disciplinary experience of having their expressions and ideas sharpened through examination by other students as well as by the teacher.

Since one can hardly present the results on *one* experiment as established conclusions, the writer suggests that the advantages of the student experience-discovery approach be investigated further. To continue the teaching of the calculus by methods which are antithetical to those of the experience-discovery approach *may* be depriving our students of a richness of mathematical education which they deserve and which it is so imperative that they have.

ANALYSIS OF CALCULUS PROBLEMS

By **HARLEY FLANDERS**

Purdue University
Lafayette, Indiana

THE student preparing for science or engineering should enter college with the technical skills needed in calculus. Particularly, he must know how to set up and solve problems and get accurate answers. Now the typical realistic calculus problem involves only a little of the ideas of calculus per se but a lot of high school algebra and geometry.

A useful way to assess the students' needs in calculus is carefully to dissect the solutions of good calculus problems into what is calculus and what is precalculus—that is, high school—mathematics. I present two such problems here, with detailed analyses of their solutions.

An isoceles triangle has area 1. How large must its perimeter be?

(1) The perimeter must be expressed as a function of the (constant) area and one tractable variable, then minimized.

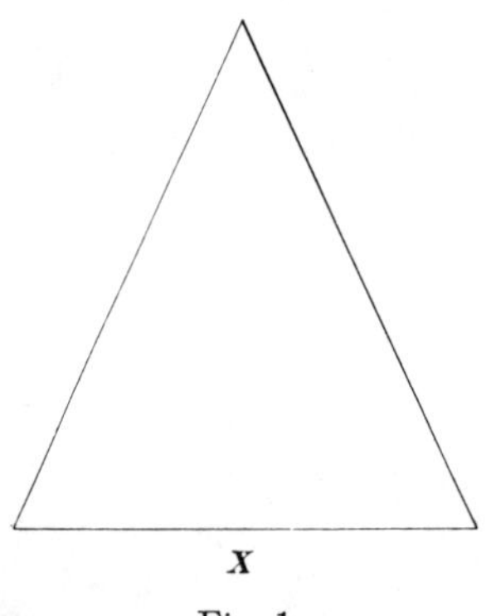

Fig. 1

[Recognition that this is a minimum problem and that it must be set up.]

(2) Draw a picture and select the variable. Maybe the base x will do the trick (fig. 1).

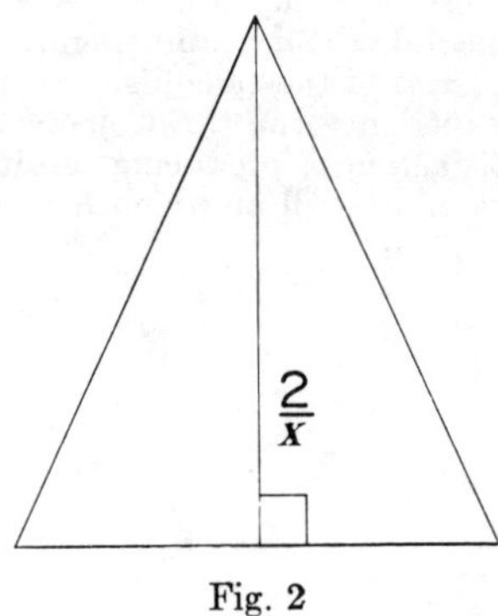

Fig. 2

(3) $A = \frac{1}{2}bh = 1$, so $h = \frac{2}{x}$ (fig. 2). Now I'm set up for finding the size z

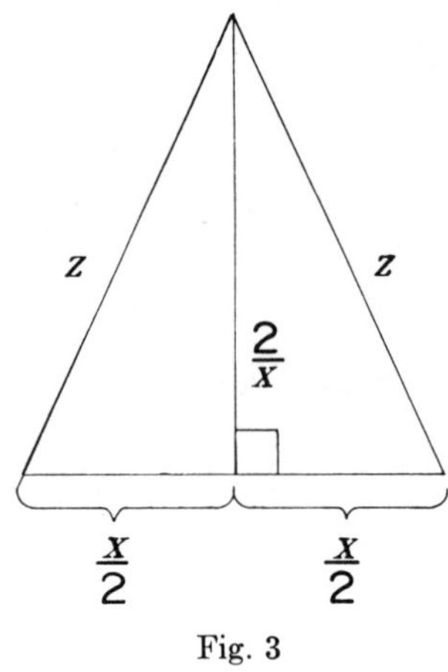

Fig. 3

by Pythagoras (fig. 3). [This step requires experience and geometric know-how.]

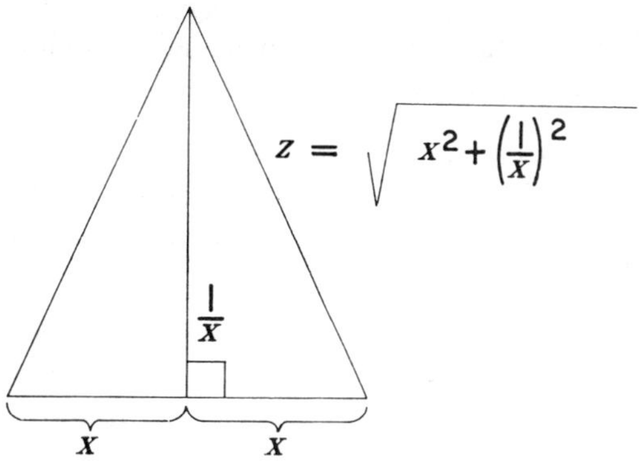

Fig. 4

(4) But those 2s are annoying. Why not call the base $2x$? (See fig. 4.) Then the semiperimeter $S = \frac{1}{2}p$ satisfies

$$S = S(x) = x + \sqrt{x^2 + \left(\frac{1}{x}\right)^2}$$

$$= x + \frac{1}{x}\sqrt{x^4 + 1}.$$

[This step uses technical skill and an appreciation of mathematical elegance.]

(5) As $x \to \infty$, then $S \to \infty$. Also, as $x \to 0+$, then $S \to \infty$. Hence, S has at least one minimum for some positive value of x.

[This is *still* high school mathematics essentially. The rough sketch (fig. 5) plus these details should suffice: for large

x we have $S \geq x$, so certainly $S \to \infty$ as $x \to \infty$; for $x > 0$ and x small,

$$S = x + \frac{1}{x}\sqrt{x^4 + 1} > \frac{1}{x}\sqrt{x^4 + 1}$$

$$> \frac{1}{x}\sqrt{1} = \frac{1}{x},$$

so $S \to \infty$ as $x \to 0+$.]

(6) Find dS/dx.

$$\frac{dS}{dx} = 1 + \frac{x\left[\dfrac{4x^3}{2\sqrt{1+x^4}}\right] - \sqrt{1+x^4}}{x^2}$$

$$= 1 + \frac{\dfrac{2x^4}{\sqrt{1+x^4}} - \sqrt{1+x^4}}{x^2}$$

$$= 1 + \frac{x^4 - 1}{x^2\sqrt{1+x^4}}$$

$$= \frac{x^2\sqrt{1+x^4} + x^4 - 1}{x^2\sqrt{1+x^4}}.$$

[Precisely the first line is calculus. The simplification is the kind of manipulative algebra the student must be able to do automatically and accurately.]

(7) Now solve $dS/dx = 0$. [This equation is calculus; the rest is algebra.]

$$x^2\sqrt{1+x^4} + x^4 - 1 = 0,$$
$$x^2\sqrt{1+x^4} = (1 - x^4),$$
$$x^4(1 + x^4) = (1 - x^4)^2,$$
$$x^4 + x^8 = 1 - 2x^4 + x^8,$$

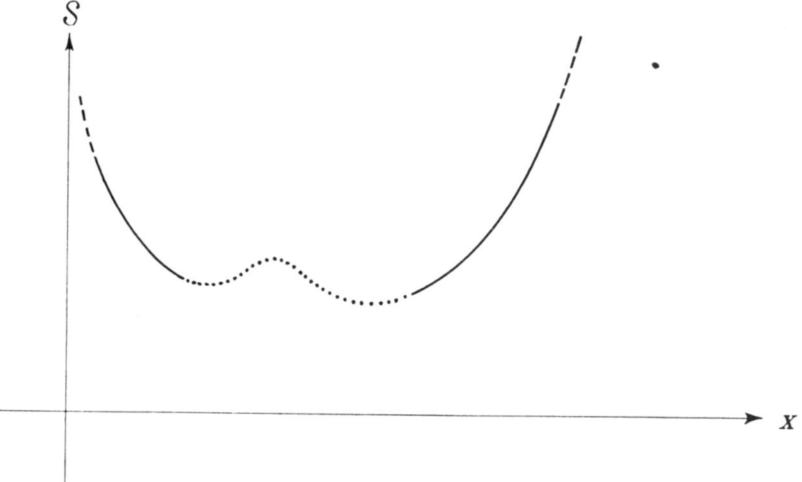

Fig. 5

$$3x^4 = 1,$$

$$x^4 = \frac{1}{3},$$

$$x = \frac{1}{\sqrt[4]{3}}.$$

[Since $dS/dx = 0$ has only one positive solution x, there is exactly one minimum.]

(8) For this $x = 1/\sqrt[4]{3}$ we have $x^4 = \frac{1}{3}$. By using *both* these relations, we can quickly calculate the corresponding value of S, then of P:

$$S = x + \frac{1}{x}\sqrt{x^4 + 1} = x + \frac{1}{x}\sqrt{\frac{4}{3}}$$

$$= x + \frac{2}{x\sqrt{3}} = x + \frac{2x^2}{x}$$

$$= 3x = \frac{3}{\sqrt[4]{3}};$$

$$P = 2S = \frac{6}{\sqrt[4]{3}} = 2\sqrt[4]{27}.$$

Answer: The smallest possible perimeter is $2\sqrt[4]{27}$. Remark: When P has this value, the triangle is equilateral because $S = 3x$, $z = 2x$.

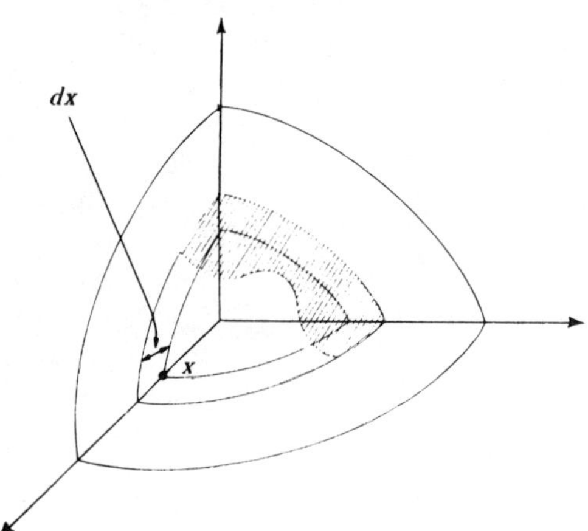

Fig. 6

[The steps in (8) are routine high school algebra. Putting down the answer to the question asked (not, for example—*Answer:* $x = 1/\sqrt[4]{3}$) is an important part of training. Recognition that the triangle is equilateral requires a good understanding of this triangle.]

The density of a material sphere decreases linearly with the central distance, from k at the center to 0 at the boundary. Find the mass of the sphere.

(1) My first step is to choose a variable, obviously the distance x from the center (fig. 6); so $0 \le x \le R$, where R is the radius.

(2) Next I must express the density function in terms of x. This is easy; just write it down:

$$\delta(x) = k(R - x).$$

This function decreases linearly to 0 at $x = R$. At $x = 0$ it is $\delta(0) = kR$.

Whoa! I'm off by a factor of R, so I'll try again:

$$\delta(x) = \frac{k}{R}(R - x).$$

That's better.

[Setting up a function to fit prescribed data is an important technique.]

(3) Now I cut the sphere into concentric shells. The mass of the shell of radius x and thickness dx is the product (density) (volume) = (density) (surface area) (thickness):

$$dM = \frac{k}{R}(R - x) \cdot 4\pi x^2 \cdot dx.$$

[This step uses elementary geometry and physics as interpreted through calculus.]

(4) Hence,

$$M = \int_0^R \frac{k}{R}(R - x) \cdot 4\pi x^2 \cdot dx$$

$$= \frac{4\pi k}{R}\int_0^R (R - x)x^2 \, dx$$

$$= \frac{4\pi k}{R}\int_0^R (Rx^2 - x^3) \, dx$$

$$= \frac{4\pi k}{R}\left[\frac{1}{3}Rx^3 - \frac{1}{4}x^4\right]\Bigg|_0^R$$

$$= \frac{4\pi k}{R}\left[\frac{1}{3}R^4 - \frac{1}{4}R^4\right]$$

$$= \frac{1}{3} \pi k R^3.$$

Answer: $M = \frac{1}{3}\pi k R^3$.

[Only the fourth line uses a technical *result* from calculus. The rest is essentially algebraic manipulation.]

These analyses of solved calculus examples suggest a method for selecting topics and placing emphasis in courses for college-bound students. Similar analyses of problems from all parts of a standard calculus course would be an excellent project for a methods course for high school mathematics teachers as well as for teachers in service, especially those considering curriculum change.

REFERENCE

Flanders, H., R. R. Korfhage, and J. J. Price. *Calculus*. New York: Academic Press, 1970.

Bibliography: *Pedagogical Overview*

Lamond, John K., C. C. Grove, and Ross W. Marriott. The Order of Teaching the Parts of the Calculus. 9 (June 1917): 191–95.

Remarks on such pedagogical questions as whether differential and integral calculus should be developed side by side.

Kinney, J. M. Calculus in the High School. 16 (October 1923): 321–31.

Remarks concerning the feasibility of teaching calculus to high school students; somewhat dated.

Cosby, Bryon. The Place of the Calculus in the Training of the High School Teacher. 16 (November 1923): 431–39.

Remarks concerning the need for a calculus requirement in the training of high school teachers; a somewhat dated presentation.

Farmer, Susie B. The Place and Teaching of Calculus in Secondary Schools. 20 (April 1927): 181–202.

Remarks concerning the feasibility of teaching calculus to high school students. Specific pedagogical suggestions for presenting basic concepts. Contains a number of technical errors.

Schaaf, William L. Calculus in the Classroom. 45 (November 1952): 525–27.

A bibliographical listing.

Kucinski, Romuald A. An Introduction to Calculus for Junior High School Students. 52 (April 1959): 250–55.

Remarks concerning the feasibility of teaching calculus to junior high school students. Some specific pedagogical approaches are presented.

Taylor, Angus E. Convention and Revolt in Mathematics. 55 (January 1962): 2–9.

Remarks on such pedagogical questions as whether calculus should be presented initially in a precise, logical fashion with no appeal to intuition or to physical ideas.

Adler, Irving. The Changes Taking Place in Mathematics. 55 (October 1962): 441–51.

An overall survey of mathematics today designed to provide insight for more effective mathematics teaching.

Nadler, Maurice. The Demise of Analytic Geometry. 62 (October 1969): 447–52.

Remarks on the current status of analytic geometry in the curriculum. Contains some examples of problems handled by both calculus and noncalculus techniques.

3

Functions

Perhaps the most frequently used mathematical term today is the word *function*. Students are currently exposed to this term as early as elementary algebra. Unfortunately, by the time they take calculus, many retain only a hazy idea of the meaning of the word.

The articles in the first part of this section deal with the general concept of function. Bennett provides a historical perspective on the evolution of the definition. Robinson extends the domain and range of a function from numbers to ordered pairs of numbers to statements.

The second part of this section is devoted to the trigonometric functions. The opening article by Boyer, meticulously researched, provides historical background on differentiating and integrating the sine function. Then Cell, in a comprehensive survey, examines eleven different approaches for defining the sine and cosine functions. The section closes with a brief but cogent reminder by Lipsey and Snow of the role that radian measure plays in the calculus of the trigonometric functions.

The exponential, hyperbolic, and logarithmic functions are discussed in the third part of this section. Bakst presents a detailed development of the hyperbolic functions, stressing the fact that the properties of these functions are analogous to those of the circular functions. Yates approaches both the logarithmic and exponential functions from the point of view of a power series. Hummel and Seebeck propose a direct postulational approach to the definition and study of logarithms. The exponential function is then defined as the inverse of the logarithmic function. In the last article, Leetch discusses an application of exponential growth.

This section reinforces the importance of the function concept as a central theme basic to mathematics. It is vital that students appreciate this early in their study of mathematics.

Concerning the function concept

ALBERT A. BENNETT, *Brown University, Providence, Rhode Island.
There are relatively recent demands for a change
in the approach to the definition of a function. This paper
recalls some of the historical background and
presents the arguments for a change.*

DURING THE LAST few decades nearly every enterprising author of a mathematics textbook designed for use in the secondary school or in the Freshman year in college has paid conspicuous homage to the function concept. The homage, even though it may have been unduly fulsome and often superficial, had been long overdue. For it has not always been thus. D. E. Smith in his remarkable and encyclopedic *History of Mathematics* (1925) devotes Volume II (725 pages) to "Special Topics of Elementary Mathematics." Although 68 major headings cover elementary mathematics of the traditional sort very thoroughly, the word "function" is not even listed in the index.

There seems no question that in perfecting drill in mathematical techniques, teachers have ignored and perhaps remained unconscious of many broad underlying ideas without which mathematics could lay little claim to its historic inclusion among the humanities as a branch of philosophy. E. H. Moore of The University of Chicago a half-century ago, in pointing out the importance of emphasizing unifying principles such as that of the function concept, succeeded in some measure in reversing the fashion. He would doubtless be disappointed could he note how mild has been the reformation in the general character of mathematics texts. However, other concepts at various logical and semantic levels have enjoyed a recent resurgence of emphasis: mathematics as a mode of communication, approximate measurements, statistical methods, mathematical logic and its special symbolism, set theory, and so forth. But let me confine my remarks to the notion of function.

No one should be surprised to hear that the notion of function has undergone considerable change in recent history. It is, however, naturally highly disturbing to find mathematicians of repute making discordant statements today as to what constitutes a suitable definition of "function." Perhaps a glance backward is in order.

"Function" seems to have been once the equivalent of the word "power" in the sense of algebra. Then the only functions of x were the integral powers x^2, x^3, etc., with the possible inclusion of x^1. Later, more complicated functions, each given by a simple algebraic formula, such as x^2-3x+2, $1/x$, $\sqrt{x^2-1}$, were accepted. Prior to the work of Descartes, little attention was paid to explicit algebraic functions as constituting a family. Of an apparently quite different sort from these algebraic functions even for Descartes, were the trigonometric functions and the then recently discovered logarithms. On the basis of Descartes' emphasis on algebraic form in what had been viewed as pure geometry, it was not many years before Newton gathered together a variety of well-known cubic curves, previously thought independent, each with its own special mechanical method of generation, and supplemented the list by classifying "all" cubic curves. But even then the equation of a conic section seems hardly to have been thought of as defining a function.

The century and a half covering the work of Euler, Laplace, Legendre, Jacobi, Abel, Gauss, and Riemann (to name a few) saw an impressive and indeed phenomenal rise in the prestige of functions of a complex variable. The earlier discovery that the logarithmic function could be defined as an indefinite integral, Euler's discovery of the identity, $e^{i\theta} = \cos\theta + i\sin\theta$, and of the extension to nonintegers of the factorial function, Gauss' study of complex numbers and of the hypergeometric functions, Jacobi's thorough development of identities and series expansions for elliptic functions, Abel's work on integrals and on solution of algebraic equations of the fifth degree, Riemann's introduction of the visually suggestive Riemann surfaces—all of these helped to suggest that the functions of one or more real variables acquire symmetry and fresh significance when the independent arguments are general complex numbers. Through the use of Taylor series, a common method was at hand for studying functions earlier generated in entirely different ways. The elementary transcendental functions become a recognized class, while elliptic functions, the Gamma function, and the potential functions of Legendre became outstanding examples of higher transcendental functions. Euler's early acceptance of implicit definitions, defined not only by algebraic but also by differential equations, made multiple-valued functions no less acceptable then single-valued functions. Later work by Lie on infinitesimal groups accepted, without qualms, conditions of differentiability. The extraordinary success of Gauss in studying differential geometry seemed added reason for thinking that differentiability is a natural property of all respectable functions. To reject this property, like rejecting the principle of excluded middle, could be expected to lead only to bizarre and fruitless disputation.

Euler may be said to have glorified the formula and to have provided by his own success an ideal for many later investigators to follow. The mere existence in the complex domain, of the derivative of a function of a complex variable at a given point, sufficed to assure that the function is analytic. The viewpoint of Euler became part of the cultural background of mathematics even after critics had pointed out its shortcomings. As evidence for this pronouncement, let me cite a 782- (large size) page volume, the second edition of which, published in 1900 by the Cambridge University Press, was entitled (on the cover) *Theory of Functions*. The author, A. R. Forsyth, remarks (p. 14): "It has been assumed that the function considered has a differential coefficient. . . . It has often been called *monogenic*, when it is necessary to assign a specific name; but for the most part we shall omit the name, the property being tacitly assumed. This is in fact done by Riemann, who calls such a dependent complex simply a *function*."

Not only were the independent variables accepted as being naturally complex variables (save where artificial restrictions to real variables are imposed for reasons of geometrical appeal or physical application), but the total domain of any variable was accepted, and usually tacitly, as the totality of values for which the formula had a meaning. Forsyth in his treatise mentioned above remarks (p. 6) in a footnote: "It is not important for the present purpose to keep in view such mathematical expressions as have intelligible meanings only when the independent variable is confined within limits."

Forsyth considers the view of defining w as a function of z, where w is obtained from z by a sequence of arithmetical operations (and is thus in a literal sense analytic) and hence in such a manner that one can compute the value of w corresponding to any given value of z. Such a definition would have been acceptable to most mathematicians of Euler's time.

Instead, Forsyth announces (p. 8): "A complex quantity w is a function of another complex quantity z, when they change together in such a manner that the value of dw/dz is independent of the value of the differential element dz."

But Euler's tradition had an abundance of sharp critics from early in the nineteenth century. Fourier in his analytical study of heat flow was one of the first to emphasize the arbitrary nature of a function of a real variable. His success and prestige served to emancipate the notion of function from the restriction that an *a priori* representation of it by a single formula is necessary. Fourier seems to have "had almost a contempt for mathematics except as a drudge of the sciences," and felt none of the inhibitions then current concerning functions. He wrote: "We can extend the same results to any functions even to those which are discontinuous and entirely arbitrary." A function no longer needed to be regarded as the embodiment of an algorithm. His boldness taught mathematicians (among his contemporaries was the great Cauchy) that intuitions are often mere prejudices and that even the "obvious" until established by proof may be false. It should have been clear that not every important elementary mathematical function is an analytic function of a complex variable. The complex conjugate, $x - iy$, of the complex variable, $z = x + iy$, is an important nonanalytic function of the complex variable z. Hence also are the absolute value and the real part of z. Euler himself introduced the totient function, $\phi(n)$, of the natural number n, $\phi(n)$ being (for $n > 1$) the number of natural numbers each less than n, and relatively prime to n. Euler never made a generalization of $\phi(n)$ to a function of a complex variable.

Of course, any definition is a matter of convention. In a proper sense no definition can be in error inherently except by being self-contradictory. A given formulation may be criticised on many grounds. It may purport to agree with other accepted formulations or to accurately reflect critical usage. In mathematics the most frequent reason for rejecting a traditional phrasing of a definition in favor of new wording seems to be that it has been too restrictive to include situations of newly recognized importance. The old definition then becomes descriptive of merely a part, historically and perhaps continuingly important, but yet not as inclusive as desired for modern research. It would seem entirely appropriate to concentrate attention at secondary-school level to functions expressible by formulas themselves algebraic, or involving at worst the elementary transcendental functions: the trigonometric, inverse trigonometric, exponential and logarithmic functions, and to consider these almost exclusively for real values of the arguments. Other functions may be regarded by secondary-school pupils as "pathological," a term used for them by many mathematicians of the last century.

But let us continue our historical resumé. In revolt against the rather blind and rampant formalism particularly of the German combinatorial school under Hindenburg, Europeans generally began a serious review of the formal methods which had proved so fruitful. Abel and Cauchy were leaders in a critical movement which received great impetus later from Weierstrass and which has almost completely changed the emphasis of mathematical research. Dirichlet in 1837 proposed the following formulation: "A variable y is a single-valued function of the variable x, in the continuous interval (a,b), when a definite value of y corresponds to each value of x such that $a \leqq x \leqq b$, no matter in what form the correspondence is specified." It may be noted that the favorite phrase "one can find" is missing. The discoverability of the associated value must be inferred, if at all, from the meaning of the term, "definite."

Dirichlet's definition remained standard for a century. It presents difficulties, however. One can ask about the meaning of the words "variable," "definite," "corresponds." There is the further semantic question as to whether a variable (here y) is itself the function concerned.

W. F. Osgood, in his *Functions of Real Variables* (1936), remarks (p. 68), "At the beginning, we spoke of y as the *function*. Thus we should say: The value of the function, $y = x^2 + 1$, when $x = 1$, is 2. This is a different meaning of the term, but no confusion of ideas arises from these two uses of the word." The question as to whether to accept many-valued functions or to insist instead on several one-valued functions, is partly a matter of taste, but also one of clear thinking. Continued experience had led many mathematicians to confine "function" to the one-valued case. One does not like to see "$\sqrt{4} = \pm 2$." It seems also desirable to reject the ambiguity which Osgood accepted and to define "function" as sharply restricted to the correspondence itself. This involves rejecting the traditional phrase "y is a function of x, etc.," in favor of "y is the value at x, of the function, etc." Such an equation as "$y = \sin x$" is entirely satisfactory, but neither y nor $\sin x$ is the function in this case. The function is the sine function, represented by "sin" and y is its value at x. There is the square function, $(x)^2$, but x^2 is a variable as is x, and writing $y = x^2$ does not make y a function. One could write more formally $y = f(x)$ and define f by "$f : x \to x^2$," read "f is the function which carries x into x^2." This alone is not enough. One must then explain what values x may take on, as for example, in "$f : x \to x^2$, (x in R)," provided "R" has been defined as the set of real numbers or rational numbers, as the case may be. The functions "$f : x \to x^2$, ($0 \leqq x \leqq 1$)" and "$f : x \to x^2$ (x, an integer)" are not to be confused.

But "correspondence" is itself a vague word, which if used, should be defined.

Any thorough-going construction of a mathematical theory must involve from the first the notions of proposition, of set, and of sequence. Is "correspondence" one of those inevitable primitive logical notions hardly explicable in terms at once more simple and more general? Why must functions always have numbers for arguments and numbers for values? In answer to queries such as these, the word "function" is being gradually dropped in some circles, in favor of the less fossilized word "mapping" (borrowed from geometry and group theory), and the case of so-called "many-valued functions," is covered very effectively by emphasizing (and of course defining) the more or less inevitable term "relation."

If A, B are sets, by "$A \times B$" (called the "Cartesian product of A by B") is meant "the set of all ordered pairs (a,b), where a is an element of A, and b of B." By "a two-term relation between variables x,y, having A,B, as respective domains," is meant "a given (non-empty) subset of $A \times B$." At first this definition seems to wreak havoc with literary usage, but one soon becomes accustomed to it and to interpreting all two-term relations in accordance with it. The domain of the two-term relation is the subset D of A, such that for each a in D, there is at least one b in B, for which (a,b) is in the relation. What identifies a function (or mapping) from among other relations is that for each a in the domain D of the function, there is exactly one b in B.

Thus a function of one argument, or a mapping, is simply a one-valued, two-term relation. The term "mapping" thus includes "functional," "projectivity," and so forth. Although the phrase "conformal mapping" is old, the general use here mentioned is very recent and may be due to van der Waerden, 1937.

Useful generalizations of the concept of function

GEORGE A. ROBINSON, *University of Illinois, Urbana, Illinois.*
*A function is a special kind of set of ordered pairs
of elements, but these elements can be numbers,
ordered pairs of numbers, or statements.*

WHEN THE CONCEPT of function[1] is defined
in a suitable manner, valuable insight may
be gained into the relationships between
some of the more elementary aspects of
mathematics and some of those aspects
generally reserved for a later point in
mathematics education as being too
sophisticated for the high school student.

It is well known that a function may be
defined as a special kind of set of ordered
pairs. (If $f(x)$ is the result of evaluating
the function f for the value x, then the
ordered pair $(x, f(x))$, and all such ordered pairs for values of x for which the
function is defined, make up a set. This
set of ordered pairs is then said to be the
function f.) In this manner, the function
concept can be built up from these two relatively more primitive concepts, set and
ordered pair. If one does this, the set of all
elements which occur as first members
(values of x) of ordered pairs in the function is commonly called the *domain* of the
function. Similarly, the set consisting of
all second members (values of $f(x)$) of
ordered pairs in the function is called the
range of the function.

If we adopt the notation $\{x/\cdots\}$,
where the three dots "$\ldots$" represents
some condition on x, to abbreviate the
phrase, "the set of all elements x such
that $\ldots$," we would write the function
commonly called "x^2" as "$\{(x, y)/x$ is a
real number and $y = x^2\}$." (Read: "the set
of all ordered pairs (x, y) such that x is a
real number and y equals x-squared.")
The domain of this function is the set
$\{x/x$ is a real number$\}$, while the range is
the set $\{y/y$ is a real number and $y \geqq 0\}$.
(See Figure 1.)

Although many people customarily
think of a function as being defined by a
simple rule such as

$$f(x) = x^2,$$

there is no reason to restrict the function
concept in this fashion. For example,
Figures 2a through 2d are graphs of per-
fectly respectable functions.

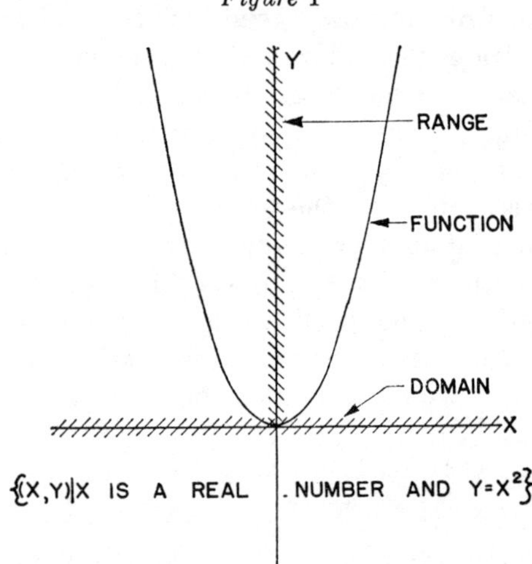

Figure 1

---

[1] In this article, the word *function* is reserved for
what is sometimes called a *single-valued function*. The
word *relation* is used to indicate a many-valued function.

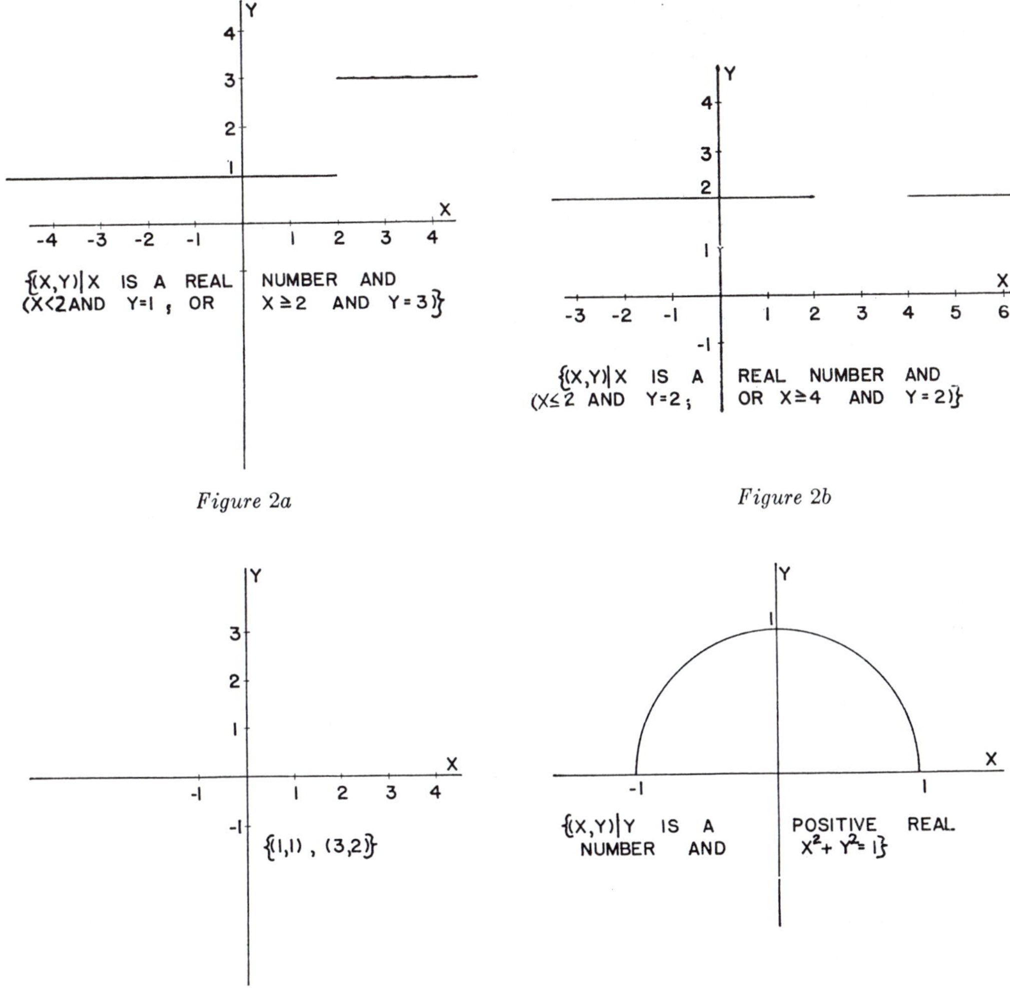

Figure 2a

Figure 2b

Figure 2c

Figure 2d

Figure 2*a* indicates that we are not requiring continuity; Figure 2*b*, that the domain need not be the entire real line; Figure 2*c*, that the function may be defined by enumerating the ordered pairs which go to make up the function; and Figure 2*d* shows that the function may be defined implicitly.

Up to this point, however, we have considered only functions whose domain and range are sets of real numbers; that is, functions which transform one real number into another real number. Nothing has been said, for example, of functions of two variables. It is not difficult to see that a function of two variables may be characterized as a function whose domain is a set of ordered pairs (of numbers) and whose range is a set of numbers. The function $f(x, y) = x^2 + y^2$ transforms ordered pairs (x, y) of numbers into the single number $x^2 + y^2$; for example, (2, 3) into 13 (i.e., into $2^2 + 3^2$); (3, 4) into 25 (i.e., into $3^2 + 4^2$); etc. Perhaps it is a little awkward, at first, to conceptualize an ordered pair, one of whose parts is also an ordered pair, for example: ((2, 3), 13).[2] but this awkwardness disappears with a little practice.

FUNDERBURG LIBRARY
c. 1
MANCHESTER COLLEGE

[2] For convenience in writing, the internal set of parentheses may be replaced by a semi-colon: (2,3;13).

TABLE 1

EXAMPLES OF FUNCTIONS WITH VARIOUS DOMAINS AND RANGES

DOMAIN	RANGE		
	Set of numbers	Set of ordered pairs of numbers	Set of statements
Set of numbers	1. ordinary functions	2. parametric representation of curve	3. condition on one variable
Set of ordered pairs of numbers	4. function of two variables	5. vector function of two variables	6. condition on two variables
Set of statements	7. truth value function	8. (no obvious useful example)	9. one-place logical operator (negation)

(It may be noted in passing that since a complex number may be thought of as an ordered pair of real numbers [i.e., $a+bi$ may be considered as the ordered pair (a, b)], a real function of a single complex variable may be considered in a manner identical to a function of two [real] variables.)

Thus we have examples of a function from a set of numbers to a set of numbers (i.e., a function with a set of numbers for its domain and a set of numbers for its range) and of a function from a set of ordered pairs of numbers to a set of numbers. We ask, "Is it possible to give a function which has a range other than a set of numbers?" This question is answered in the affirmative by Table 1, which gives examples of functions with three different kinds of sets used as domain and range.

We have already discussed cells 1 and 4 of Table 1. Before considering functions involving statements, we will consider cells 2 and 5, which deal with functions whose domains are similar to those of the two kinds of functions already considered, but whose ranges are sets of ordered pairs of numbers, rather than sets of numbers. As an example of cell 2, the function

$$\{(t, (x, y))/x = \cos t, y = \sin t\}$$

is often used to describe the unit circle. If we call this function "f," we find that

$f(0) = (1, 0)$, $f(\pi/4) = (\frac{1}{2}\sqrt{2}, \frac{1}{2}\sqrt{2})$, $f(\pi/2) = (0, 1)$, $f(\pi) = (-1, 0)$, $f(3\pi/2) = (0, -1)$, and $f(2\pi) = (1, 0)$. Thus, as t goes from 0 to 2π, the point $f(t)$ (i.e., the ordered pair (x, y), where $x = \cos t$ and $y = \sin t$) describes a circle (in the counterclockwise direction) of unit radius about the origin as a center. Such parametric representations of a curve are an important device in beginning calculus.

Moving to cell 5, we note that a vector function of two variables may be characterized as a function from a set of ordered pairs of numbers to a set of ordered pairs of numbers. For example, the wind magnitude and direction at each point of the earth's surface may be so represented as a vector function of two variables. This description of a vector function in terms of ordered pairs can become an important tool in cutting through the fog that enshrouds this topic (vector) in the minds of many of us and of our students. This application to vectors, vector functions, and the related concept of tensor forms an interesting discussion in itself, but we are limited here to noting that cell 5 is filled.

Again, in passing, we note that the "ordered-pairness" of complex numbers enables us to write a complex function of a single complex variable in a manner identical to that of a vector function of two real variables.

Another useful generalization of the function concept is obtained (cell 3) by considering a function from a set of numbers to a set of statements. Many students are confused to some extent by the nature of expressions of the form $f(x) = g(x)$, where f and g are functions. Take for example the case where $f(x)$ is $x^2 + x$ and $g(x)$ is identically zero. We are told that this expression becomes a statement (true or false) whenever the variable x is replaced by one of its possible values. When asked the question, "What is '$x^2 + x = 0$' before any substitution is made for 'x'?", one might be tempted to reject the question as improperly posed, or to answer merely by giving a name ("equation"?) to the entity. Indeed, this is the most common reaction. If, however, we remember the treatment of function formerly used in high-school texts, we would see that much the same reply was given to the question, "What is a function?" Not equipped with the concept of function in the modern sense, reference would be made to what the function (or expression) x^2 becomes when the variable "x" is replaced by each of a set of possible values. The result of each substitution in the function case was a number. For example, for the function x^2, Figure 3 gives the substitution results for a few cases.

The similarity between the equation case and the function case is so striking that we are tempted to ask the question: "Can an equation in one variable, such as $2x^2 - x = x^3$, be thought of as a function of some sort, and if so, what is the domain of this function and the range of this function?" As an aid to visualization, we may compile a table (Fig. 4) similar to the one prepared (Fig. 3) for the function case.

Figure 3

x	x^2
2	4
2.5	6.25
3	9
4	16

x	$2x^2 - x = x$
-1	$3 = -1$
0	$0 = 0$
1	$1 = 1$
2	$6 = 8$
3	$15 = 27$
4	$28 = 64$

Figure 4

We see that an equation in one variable becomes a statement[3] for each value of the variable. For example, when $x = -1$, we get the (false) statement $3 = -1$, etc. In this sense, an equation in one variable is a function with a set of numbers as its domain and a set of statements as its range.

The next logical step is to ask whether anything more general than an equation in one variable may be written as such a function. It turns out that any condition[4] on a single variable may be written as a function from a set of numbers to a set of statements. This fact is noted in cell 3. As another special case of condition, any inequality in one variable may be represented as such a function.

Cell 6 bears the same relation to cell 3 as does cell 4 to cell 1 and requires, perhaps, no special discussion. An example of a condition in two variables might be $x = y$. This condition is not to be confused with the *relation* $x = y$,

$$\{(x, y)/x = y\},$$

which is a set of ordered pairs of numbers. (This latter relation also happens to be a function, since to each x there corresponds only one y.) The condition $x = y$ is of the form

$$\{((x, y), A)/ A \text{ is a statement expressing the equality of } x \text{ and } y\}.$$

---

[3] A statement has the property of being either true or false. An equation in one variable cannot be said to have either of these properties until a value is inserted for the variable.

[4] A condition may be defined as an expression containing a variable such that when a numerical value is substituted for the variable, a statement results. Thus an equation in one variable is a condition on one variable. The term "propositional function" is frequently used in place of the other.

If we use the word *solution* in its usual sense, we might want to remark that the relation $x=y$ is the solution of the condition $x=y$.

As an example for cell 7, we construct a truth value function, f, which behaves as follows: if A is a true statement, then $f(A)=1$; if A is a false statement, then $f(A)=0$. This function could be an interesting conceptual device in developing the study of logic on the secondary level.

For cell 8, no useful example stands out to the author, although artificial examples may be constructed. Perhaps the reader can suggest a useful function qualifying for cell 8.

For cell 9, we need a function which will, when a statement is "plugged in," give a statement as the result. The negation operator of symbolic logic is such a function:
$$\{(A, B)/A \text{ and } B \text{ are statements},$$
$$A \text{ is not } -B\}.$$

For example, if we let "$\sim$" denote the above function:

$$\sim(3=6)=(3\neq6)$$

$$\sim(3\neq6)=(3=6)$$

$$\sim(\text{it is raining})=\text{it is not raining}.$$

In conclusion, we should note that, in setting up Table 1, the choice of numbers, ordered pairs, and statements for discussion is not the only choice we might have made. Columns and rows referring to functions with sets of sets and sets of functions as domain and/or range might have been included. The former is of fundamental importance in theory of probability and mathematical statistics, while the latter finds use in advanced problems in electrical engineering. The reader may be able to think of still other profitable choices for columns and rows.

Bibliography: *Concepts and Notation of Functions*

Webb, Harrison E. Professor Hedrick's Report on the Function Concept in Elementary Mathematics. 15 (October 1922): 364–68.

Somewhat dated remarks on the importance of the function concept in elementary mathematics.

Kinney, J. M. The Function Concept in High School Mathematics. 15 (December 1922): 484–95.

Somewhat dated comments; specific suggestions on the restructuring of high school mathematics courses around functionality.

Cronbach, Lee J. What the Word "Function" Means to Algebra Teachers. 36 (May 1943): 212–18.

Remarks concerning widespread disagreement among algebra teachers on the precise definition of function.

Rich, Barnett. The Place of the Variable in the Teaching of Mathematics. 48 (December 1955): 538–41.

Pedagogical suggestions for presenting the concept of variable.

Wollan, G. N. What Is a Function? 53 (February 1960): 96–101.

Remarks on the definition and meaning of function.

Wampler, J. F. The Concept of Function. 53 (November 1960): 581–83.

Remarks on the definition and meaning of function.

Hight, Donald W. Functions: Dependent Variables to Fickle Pickers. 61 (October 1968): 575–79.

Remarks on the historical development of the function concept and a student-centered definition of the term.

Leetch, J. F. A Dialogue on Inverse Functions. 63 (November 1970): 563–65.

A pedagogical approach to the study of inverse functions.

Brieske, Thomas J. Functions, Mappings, and Mapping Diagrams. 66 (May 1973): 463–68.

Pedagogical suggestions for presenting the function concept.

History of the Derivative and Integral of the Sine

By CARL B. BOYER
Brooklyn College, Brooklyn, N. Y.

THE early development of the sine function was geometric rather than analytic. In ancient Greece the function concept and the ideas of differentiation and integration were not given formal expression, but the equivalents of these notions were implied by a number of problems in geometry. The determination of the length of a chord in a circle subtending a given angle x, for example, was the Greek equivalent of finding the value of the function $2 \sin x/2$. The method of exhaustion was the Greek geometric counterpart of finding an area by integration; and the kinematic determination by Archimedes of the tangent to his spiral is tantamount to the differentiation of a function. The calculus of the sine, however, could not at that time be given in terms of the now customary area and slope interpretations inasmuch as the sine curve was then unknown. Nevertheless, there are in the works of Archimedes several propositions analogous to the determination of the integral of the sine.

Preliminary to finding the surface and volume of a spherical segment, Archimedes proved[1] the following proposition:

Let AA' be the diameter of a circle and let QAQ' be any arc bisected at A. Let this arc be subdivided into $2n$ equal parts by the points $A, B, B', C, C', \cdots, P, P',$ Q, Q' (Fig. 1). Then

$$BB'+CC'+ \cdots +PP'+QM):AM$$
$$=A'B:BA.$$

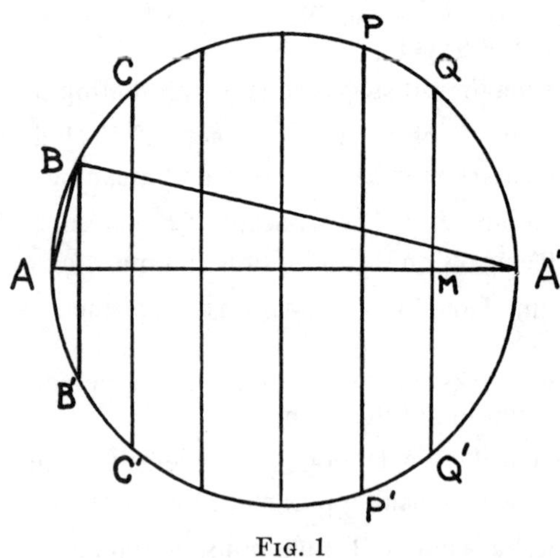

FIG. 1

---

[1] *On the Sphere and Cylinder* I, Proposition 22. See *The Works of Archimedes* (ed. by T. L. Heath, Cambridge, 1897), pp. 28–29; cf. pp. cxliv–cxlvi. For other aspects of the trigonometry of Archimedes see Johannes Tropfke, "Archimedes und die Trigonometrie," *Archiv für die Geschichte der Mathematik, der Wissenschaften, und der Technik*, X (1927–1928), 432–463.

This is the geometrical equivalent of the trigonometric equation

$$\sin\frac{\theta}{n}+\sin\frac{2\theta}{n}+\cdots+\sin\frac{(n-1)\theta}{n}$$

$$+\frac{1}{2}\sin\frac{n\theta}{n}=\frac{1-\cos\theta}{2}\cot\frac{\theta}{2n}\cdot$$

From this theorem of Archimedes it is a simple matter analytically to derive the modern expression $\int_0^\theta \sin x\,dx = 1-\cos\theta$ by multiplying both sides of the trigonometric equation above by θ/n and taking limits as n increases indefinitely. The left-hand side becomes

$$\lim_{n\to\infty}\sum (\sin x_i)\Delta x_i$$

$$\left(\text{where } x_i=\frac{i\theta}{n}\text{ for } i=1,\ 2,\ \cdots,\ n,\text{ and}\right.$$

$$\Delta x_i=\frac{\theta}{n}\text{ for } i=1,\ 2,\ \cdots,\ n-1$$

$$\left.\text{and } \Delta x_n=\frac{\theta}{2n}\right);$$

the right-hand side becomes

$$(1-\cos\theta)\lim_{n\to\infty}\left(\frac{\theta}{2n}\cot\frac{\theta}{2n}\right)=1-\cos\theta.$$

The equivalent of the special case

$$\int_0^\pi \sin x\,dx = 1-\cos\pi = 2$$

had been given by Archimedes in an earlier theorem. The language of Archimedes in his quadratures and cubatures follows the usual pattern of the ancient method of exhaustion, with both inscribed and circumscribed figures and with an argument by a reductio ad absurdum in lieu of the passage to the limit; but the fundamental notions are quite analogous to those formulated in the modern symbolism of the definite integral.

The infinitesimal methods of Archimedes were continued in antiquity by the Alexandrian mathematicians, including Pappus, and afterwards passed over into the Arabic civilization.[2] Meanwhile, the

Greek trigonometry of chords, developed by Hipparchus and Ptolemy, was converted by the Hindus into the study of half-chords or sines, and this, too, was passed on to the Arabs. Through the Muslim civilization this knowledge was in turn transmitted to the Latin world, especially during the twelfth and thirteenth centuries. By the sixteenth century the works of Archimedes were well known in Europe, and one finds Cardan going over a problem on the circle similar to the ancient geometrical integration of the sine.[3] Early in the following century Kepler cited Cardan in this connection and used an arithmetical approximation to this integral in his *Astronomia nova* of 1609. In his calculations on the planet Mars he had occasion to add the sines of angles, for every degree, from 1° to 15°. This sum he found to be 2.08166, which when multiplied by an approximate value for $\pi/180$ gave him .03594. This, he noted, is a little more than the versed sine of 15°, which he took as .03407. Similarly he found the sum for 30° to be 7.92598, which, when multiplied as above "by the rule of proportional parts," becomes .13691, or a little more than vers 30° = .13397. Kepler found a similar agreement for 60° and 90° also.[4] Such calculations are similar to determinations of

$$\sum_{i=1}^{n} (\sin x_i)\Delta x_i,$$

where $x_i = i°$ and $\Delta x_i = \pi/180$, and hence are approximations to $\int_0^\theta \sin x\,dx = 1-\cos\theta$.

---

[2] See Heinrich Wieleitner, "Das Fortleben der Archimedischen Infinitesimalmethoden bis zum Beginn des 17. Jahrhundert," *Quellen und Studien zur Geschichte der Mathematik, Astronomie und Physik, Part B, Studien,* I (1931), 201–220.

[3] *De subtilitate,* XVI. See Hieronymus Cardanus, *Opera omnia* (10 vols., Lugduni, 1663), III, 593. This passage is given in Johann Kepler, *Opera omnia* (ed. by Ch. Frisch, 8 vols., Frankoforti A. M. & Erlangae, 1858–1870), III, p. 497, note 86. Citations of Kepler's work given in this paper are all based on this edition. Several volumes of a new edition of Kepler's works by Max Caspar have appeared recently.

[4] *Opera omnia,* III, 390–391. Cf. also pp. 105 and 335. For purposes of exposition the lan-

They thus represent an arithmetization of the geometrical proposition of Archimedes. However, whereas Archimedes had given, through the method of exhaustion, the equivalent of the *limit* of a sum, Kepler did not here link his result with limits or infinitesimals, and it is not clear just what significance he attached to his approximation.[5] Shortly afterwards, in attempting to calculate the magnitude of the attraction between the sun and the earth, he sug-

Inspired largely by the works of Archimedes and Kepler, mathematicians of the second third of the seventeenth century rapidly expanded infinitesimal geometry until it culminated in the calculus. It was one of the leaders in this movement who, about 1635, took the next step in the integration of the sine. Roberval first reproduced the proposition of Archimedes on the sum of chords in a circle, putting it in the then current terminology of sines and

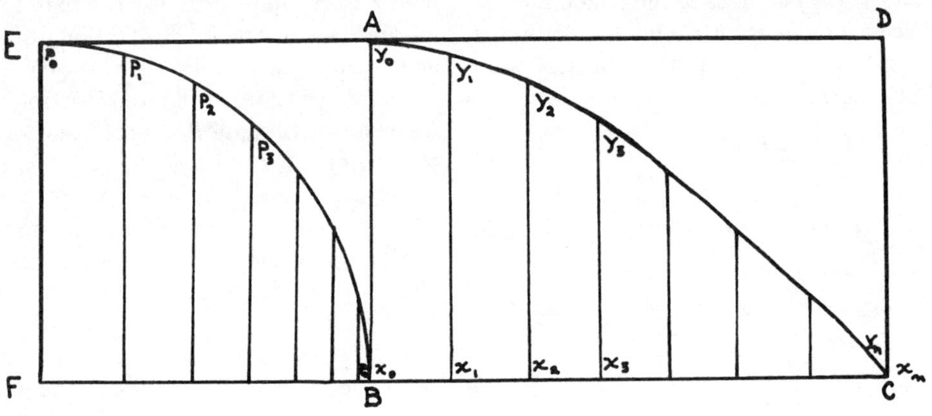

Fig. 2

gested a corresponding geometrical demonstration, in terms of infinitesimals, along the lines of one given by Pappus on volumes of spherical segments.[6]

infinitesimals: "If any arc of a circle is divided into an infinite number of equal parts and if it is projected orthogonally upon a diameter, the projection is to the arc as the sum of all the sines drawn from the points of division is to the product of the radius and the arc." The proof of this proposition, which is geometrically equivalent to $\int_{\theta_1}^{\theta_2} \sin x\,dx = \cos\theta_2 - \cos\theta_1$, is along the general lines of that given by Archimedes; but it differs in that the number of divisions is taken as infinite and an infinitesimal arc is substituted for an infinitesimal chord.[7] A corollary to this proposition gives the special case equivalent to $\int_0^{\pi/2} \sin x\,dx = 1$.

Roberval then gave a strikingly original proposition in which a portion of the sine curve appeared for the first time in his-

---

guage and notation of Kepler have been somewhat modified. Sines were lines in Kepler's day, and the ratio definitions of the trigonometric functions were not systematically used until the time of Euler. For a good general account of Kepler as a mathematician see Fritz Kubach, "Johannes Kepler als Mathematiker," *Veröffentlichungen der Badischen Sternwarte zu Heidelberg*, vol. XI, 1935. A briefer account is found in *D. J.* Struik, "Kepler as a mathematician," in *Johann Kepler, 1571–1630* (Baltimore, 1931), pp. 39–57

[5] For a full analysis with references see Gustav Eneström, "Ueber die angebliche Integration einer trigonometrischen Function bei Kepler," *Bibliotheca Mathematica* (3), XIII (1912–1913), 229–241.

[6] *Opera omnia*, VI, 407. See Siegmund Günther, "Über eine merkwürdige Beziehung zwischen Pappus und Kepler," *Bibliotheca Mathematica*, new series, II (1888), 81–87; Gustav Eneström, "Sur un théorème de Kepler équivalent a l'intégration d'une fonction trigonométrique," *Bibliotheca Mathematica* (new series), III (1889), 65–66.

[7] See Evelyn Walker, *A Study of the Traité des Indivisibles of Roberval* (New York, 1932), pp. 79–81, 178–179; or G. P. de Roberval, "Divers ouvrages," *Mémoires de l'Académie Royale des Sciences depuis 1666 jusqu'à 1699*, VI (Paris, 1730), 1–478.

tory. Letting the line BC (Fig. 2) be equal to the quadrantal arc EB of a circle of radius $r = AB = FB$, he subdivided the line and the arc into n equal parts by points x_0, x_1, x_2, x_3, $\cdots$, x_n and P_0, P_1, P_2, P_3, $\cdots$, P_n. Then at each of the points x_i he erected an ordinate $x_i y_i$ equal in length to the perpendicular from P_i to the line FBC. Plotting the locus of the points y_i he obtained a curve which at the time was known as "Roberval's curve." It is, of course, half of one arch of the now familiar curve of sines.[8] Roberval then showed that the area under this curve is equal to r^2. This is easily deduced from the corollary above, for

$$AB : BC = \sum \overline{x_i y_i} : AB \cdot BC.$$

Clearing of fractions, the result follows from the fact that $\sum \overline{x_i y_i}$ is the area under the curve. Here one finds for the first time the interpretation of the definite integral of the sine in terms of the area under the sine curve; but it was more than a century before this became customary.

In 1659, just about a quarter of a century after Roberval's treatment of the sine, there appeared two small works devoted primarily to the integration of this function. One of these—the *Opusculum geometricum de linea sinuum et cycloide* of Honoré Fabri—illustrated the newer approach of Roberval; the other—the *Traité des sinus du quart de cercle* of Blaise Pascal—continued the older tradition of Archimedes. Referring only to a "certain anonymous celebrated geometer" as his inspiration, Fabri determined the area and center of gravity of the "figure of sines," as well as the volumes, surface areas, and centers of gravity of solids of revolution associated with the curve. Then he solved similar problems in connection with the cycloid, work that was overshadowed by Pascal's famous *Histoire de la roulette*.

In the *Traité des sinus* Pascal first re-

stated the Archimedean result—"The sum of sines of any arc of a quarter of a circle is equal to that portion of the base included between the extreme sines, multiplied by the radius."[9] This theorem, equivalent to

$$\lim_{n \to \infty} \sum_{i=1}^{n} \sin x_i \Delta x_i = \cos \theta_2 - \cos \theta_1,$$

where $x_0 = \theta_1$ and $x_n = \theta_2$, he easily proved in connection with a diagram (Fig. 3) which later became famous through the work of Leibniz. From the similarity of the right triangles EKE and DIA, and the resulting equality $DI \times EE = RR \times AB$, the

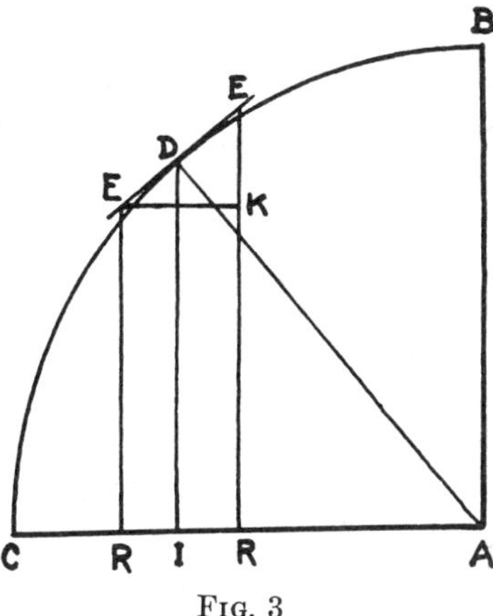

Fig. 3

theorem followed on making the intervals RR infinitely small and on substituting for the infinitesimal tangent line segments EE the infinitesimal arcs cut off by the ordinates ER. Pascal then went beyond the Archimedean result and showed similarly that the sum of the nth powers of the sines is equal to the radius times the sum of the $(n-1)$st powers of the ordinates. The theorem is equivalent to the equation

$$r^{n+1} \int_{\theta_1}^{\theta_2} \sin^n x \, dx = r \int_{r \cos \theta_1}^{r \cos \theta_2} y^{n-1} dx.$$

It is to be remarked that Fabri and Pascal considered only the definite inte-

<hr>

[8] Walker, *op. cit.*, pp. 66–68, 180; Roberval, *op. cit.*, pp. 293–345. Roberval's diagram and notation have been somewhat modified for greater clarity.

[9] Blaise Pascal, *Lettres de A. Dettonville* (Paris, 1659), p. 1 f; or see his *Oeuvres* (ed. by Brunschvicg and Boutroux, 14 vols., Paris, 1908–1914), IX, 60–76.

gral. The indefinite integral or anti-derivative requires for its definition the formulation of the derivative, the basis for which was being laid at the very time these men were writing—and in connection with which Pascal played an inadvertent role. Their contemporary, Fermat, had developed an ingenious method of determining maxima and minima of a function $f(x)$. In this he first wrote $f(x+E)-f(x)$, and then, after simplifying by division and otherwise, he dropped out those terms which still contained E. This process is equivalent to finding

$$\lim_{E \to 0} \frac{f(x+E)-f(x)}{E}$$

and then equating this to zero. Fermat's method, which he applied also to the determination of tangents, is as closely analogous to differentiation as is the work of Archimedes to integration. Fermat and his contemporaries, however, seem to have been concerned largely with algebraic functions, and so the method was not applied directly to Roberval's curve of sines. The differential calculus of the sine function arose somewhat as had the integral long before—through the geometry of the circle.

Leibniz tells us that it was through a study of Pascal's work that he was led to his differential calculus.[10] Apparently it was in connection with the diagram for the *Traité des sinus* (Fig. 3 above) that Leibniz first realized that Fermat's method of differences was simply the determination of the ratio of two sides of the infinitesimal triangle EKE. Letting $RR = dx$ and $EK = dy$, the slope of the tangent at any point is thus given by dy/dx, or the ratio

of two differentials. The differential or characteristic triangle played a prominent role in the calculus of Leibniz, and it was from this, through the obvious relationship $ds^2 = dx^2 + dy^2$, that the derivative of the sine arose. Although transcendental functions were not explicitly included in the early expositions published by Leibniz and Newton, it appears that the differential or fluxion of the sine was known from the earliest days of the calculus. A manuscript by Leibniz of 1676 includes the statement[11] that in a circle of radius r, if z is the arc of which x is the sine of the complement, and if in place of x one takes $x + \beta$ and in place of z one takes $z - dz$, then $dz = \dfrac{\beta r}{\sqrt{r^2 - x^2}}$. This expression for dz is equivalent to $d \arccos \dfrac{x}{r} = \dfrac{-r dx}{\sqrt{r^2 - x^2}}$.

Inasmuch as the characteristic triangle formed the basis of the differential method of Leibniz, it is probable that the result above, as well as the differential of the arcsine, was derived from the geometry of the circle. For any point P on a circle (Fig. 4) one deduces immediately, from the similarity of the triangles with sides dx, dy, dz and x, y, r, the results

$$dz = \frac{-r dx}{\sqrt{r^2 - x^2}} \text{ and } dz = \frac{r dy}{\sqrt{r^2 - y^2}}, \text{ where}$$

dz is $d \arccos x/r$ or $d \arcsin y/r$. In published works of Leibniz and the Bernoullis,[12] such diagrams and relationships appear quite incidentally, presumably because they were well known. There was at the time no clear distinction between dependent and independent variables, and so these equations appeared in various forms. The relationship $r^2 dz^2 = r^2 dy^2 + y^2 dz^2$, for example, is easily converted into the inverse trigonometric form given above;

<hr>

[10] See Dietrich Mahnke, "Neue Einblicke in die Entdeckungsgeschichte der höheren Analysis," *Abhandlungen der Preussische Akademie der Wissenschaften*, Physikalisch-Mathematische Klasse, I (1925), 1–64. Child, however, believes that Leibniz was led to his calculus by the work of Barrow rather than Pascal. See *The Early Mathematical Manuscripts of Leibniz* (transl. with notes by J. M. Child, Chicago and London, 1920).

[11] Leibniz (Child), *op. cit.*, p. 116 f.

[12] See *Acta Eruditorum*, III (1686), 297; VII (1691), 287, and Tab. VIII, fig. IV; X (1693) 179; XI (1694), 395–396, and Tab. XI, fig. 4. These are cited also in Eneström, "Die erste Herleitung."

or, on writing it as $d\left(\dfrac{y}{r}\right) = \dfrac{\sqrt{r^2-y^2}}{r^2}\, dz,$

one may express it as $d \sin z = 1/r \cos z\,dz$ (where z is the arc) or as $d \sin \theta = \cos \theta\,d\theta$ (where θ is the central angle).

The calculus of the sine and arcsine were known also to Newton from the early days of his method of fluxions. He saw

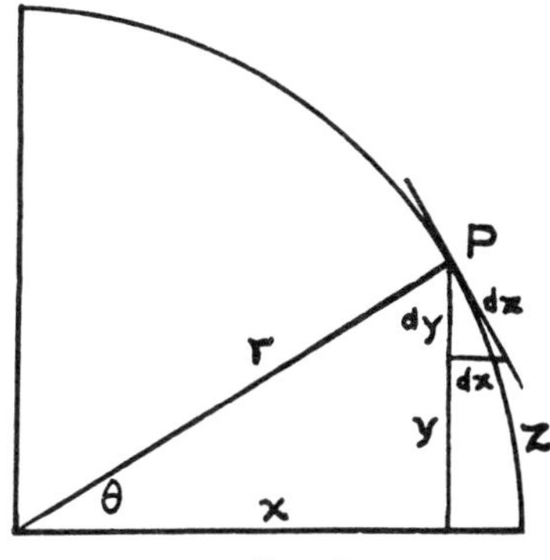

FIG. 4

likewise, through the characteristic triangle and the geometry of the circle, that

$\dot z = \dfrac{r\dot y}{\sqrt{r^2-y^2}},$ where $\dot z$ is the fluxion of the arc corresponding to the fluxion $\dot y$ of the sine of the arc. (Fig. 4.) He then converted $\dfrac{1}{\sqrt{r^2-y^2}}$ into an infinite power series in y, using the binomial theorem.[13] On integrating this term by term, he obtained the series for arcsin y; and through his method for the reversion of series, the expansion for sin z was obtained. Using the relationship $r^2 = x^2 + y^2$, it was a simple matter for him to find the expansion for cos z by the binomial theorem. On applying the method of fluxions to the terms of these series, it was clear that the fluxion of the sine is given by the cosine, and that the fluxion of the cosine is the negative of the sine. These results were communicated to Oldenburg in 1676, the year of the Leibniz manuscript mentioned above, but Newton seems to have discovered them

some years earlier.[14] James Gregory also arrived independently at these results at about the same time and in much the same manner,[15] and shortly afterwards Leibniz too was familiar with these series. It is interesting to note that whereas such series now are invariably derived through successive differentiation, they were obtained at that time through integration.

During the latter part of the seventeenth century infinite series played an increasingly important role in analysis, especially in connection with the trigonometric functions. Nevertheless, the trigonometry of the time remained a loosely organized study of geometrical propositions on lines in a circle. The ratio definitions had not yet entered; the trigonometric curves were scarcely known; and symbolisms, other than the conventional designation of lines by letters, were rarely used. It is not surprising, therefore, that the calculus of these functions was not at the time systematized. Inasmuch as trigonometric functions did not appear explicitly in equations, rules of differentiation and integration were not formally given. Where necessary, the fluxions or differentials of trigonometric lines were easily derived from the characteristic triangle or from the known infinite series representations. Moreover, trigonometric lines were regarded as functions of the arc rather than of the central angle, and hence the differentials of these were derived generally from the differentials of the inverse functions.

During the first half of the eighteenth century the role of the circular functions in the calculus was not greatly changed, but a new element, the idea of periodicity, became more persistent in trigonometry. The periodicity of the trigonometric functions generally has been ascribed to DeLagny who in 1705 indicated this for the tangent, but it is now known that the idea is at least thirty-five years older. The

<hr>

[13] Newton, *Opera omnia* (ed. by Samuel Horsley, 5 vols., Londini, 1779–1785), I, 297 f.

[14] *Opera omnia*, I, 297 *f*, and IV, 451.
[15] James Gregory, *Tercentenary Memorial Volume* (London, 1939), pp. 6, 61, 155.

first clear-cut recognition of trigonometric periodicities seems to be that of Wallis in the *Mechanica* of 1670. Here two full cycles of the sine curve are clearly drawn; and the periodicity is evidenced by an accompanying statement that the sine curve increases throughout the first, fourth, fifth, eighth, ninth quadrants, *and so forth*.[16] Wallis seems to have been led to the idea of periodicity from the relation of the sine curve to the cycloid, where the repetition of cycles is more obvious to the eye from the manner in which it is generated. However, the periodicity of Wallis did not make a deep impression. Only much later, when the analytic goniometry of Euler showed that the multiple-angle formulas corroborated the idea, did the periodicity of the functions find general acceptance.

The brilliant young mathematician Roger Cotes was one of the few who in 1722 recognized the periodicity of the trigonometric functions, for he drew cycles of the tangent and secant curves; but his calculus of the functions remained essentially that of the previous century. He formalized the geometrical calculus of the cyclometric functions in the following lemma: "The least variation [or differential] of any circular arc is to the least variation of the sine of this arc as the radius is to the sine of the complement."[17] This is equivalent to the familiar result $d \sin \theta = \cos \theta d\theta$, where θ is the central angle corresponding to the circular arc whose sine is y. This is the clearest and most explicit statement up to that time on the differentiation of the sine, but the justification was along the traditional geometrical lines of Leibniz and the Bernoullis given above. Again in 1742, in Maclaurin's well known *Treatise of flux-*

ions, formal rules were given for powers, quotients, and logarithms, but the fluxions of the circular functions were still dependent upon the geometry of the characteristic triangle.[18]

The modern presentation of the trigonometric functions and their derivatives stems largely from the work of Euler. In his *Introductio in analysin infinitorum* of 1748 one finds for the first time a truly functional treatment of trigonometry. Influenced perhaps by the infinite series representations, Euler emphasized the numerical or analytic aspect of the subject, so that here the trigonometric functions are defined as ratios rather than lines. Clear graphs of the functions are drawn, showing their periodicities; and all the customary formulas of goniometry are given in terms of the now usual symbolic abbreviations.[19] However, Euler still defined the trigonometric functions in terms of the arc rather than the central angle.

During the half century following 1696 a number of textbooks on the differential calculus had appeared, but they did not include the trigonometric functions as such. The formal analytic treatment of the calculus of the sine begins with Euler's *Institutiones calculi differentialis* of 1755. Here the differential of the arcsine is first derived from that of the logarithm by means of the complex relationship

$$y \,=\, \text{arc}\,\sin x = \frac{1}{i} \log \left(\sqrt{1-x^2} + ix \right).$$

This represents a continuation of the tradition which made the trigonometric functions subsidiary to others and which placed the inverse sine before the sine. Euler then added, however, that one can derive the differential of the arcsine more easily without logarithms as follows:[20]

[16] John Wallis, *Opera* (3 vols., Oxonii, 1693–1699), I, 504–505 and figure 201 opposite page 542.

[17] Cotes, *Aestimatio errorum*, p. 3. This is bound with his *Harmonia mensurarum* (Cantabrigiae, 1722), but is separately paginated. See *Harmonia mensurarum*, pp. 78–81, for the tangent and secant curves.

[18] Colin Maclaurin, *Traité des fluxions* (transl. by the R. P. Pezenas, 2 vols., Paris, 1749), I, 126 f.

[19] Leonhard Euler, *Introductio in analysin infinitorum* (2 vols., Lausannae, 1748). See especially I, 93 ff, and II, fig. 104.

[20] *Opera omnia* (23 vols., Lipsiae, 1911–1938), series 1, vol. X, pp. 132 ff. The deriva-

If $y = \arcsin x$, then $x = \sin y$ and $x + dx = \sin (y + dy)$. But $\sin (y + dy) = \sin y \cos dy + \cos y \sin dy$, and for an "evanescent" arc the sine may be taken as the arc and the cosine may be taken as unity. Hence $x + dx = \sin y + dy \cos y$, or $dx = \cos y\, dy$, or

$$dx = dy \sqrt{1 - x^2}, \quad \text{or} \quad dy = \frac{dx}{\sqrt{1 - x^2}}, \quad \text{or}$$

$$dy = \frac{dx}{\cos y}.$$

From the above it is obvious that the differential of $\sin y$ is $\cos y\, dy$; but before pointing this out, Euler preferred to derive "the inverse of the arcsine" directly by repeating the above argument for $y = \sin x$. This fact represents perhaps an early appreciation of the need for distinguishing between dependent and independent variables. He now wrote $y + dy = \sin (x + dx)$ or $dy = \sin (x + dx) - \sin x$, and for $\sin (x + dx)$ substituted $\sin x \cos dx + \cos x \sin dx$. Euler then replaced $\sin dx$ by dx and $\cos dx$ by 1, justifying this here in terms of the series expansions of the functions, and obtained $dy = \cos x\, dx$.

The treatises and papers of Euler exerted a decisive influence, and the trigonometric functions appeared regularly in calculus textbooks of the second half of the century. The Abbé Sauri in his *Cours complet de mathématiques* made special reference in an historical introduction to the fact that "Euler had added to analysis the new calculus of the sine and cosine"; and in the volume on the differential calculus he gave the Eulerian treatment of these functions.[21] When in the following year a French edition of the successful Italian calculus text of Maria Agnesi appeared,

the editor added in an appendix the differentials of the sine and cosine, derived in the manner of Euler.[22] A half dozen years later a revised edition of the ever popular text of L'Hospital included an extensive note on the differential and integral of the sine and arcsine, derived geometrically first, and then analytically through Euler's identities.[23]

At the turn of the century the Eulerian calculus of the trigonometric functions was popularized by the ubiquitous textbooks of Lacroix; but even as these were appearing there was an insistent demand for a more rigorous treatment of analysis, free from Euler's infinitesimal zeroes. The wide search for a firm foundation led finally to the limit concept as the basis both in differentiation and in integration. Such an approach had been suggested by Newton, was more definitely indicated by D'Alembert and L'Huilier, and was made rigorous by Bolzano and Cauchy. It was the classic treatises of Cauchy in particular which led to the triumph of the limit concept, and in his *Résumé des leçons sur le calcul infinitésimal*[24] of 1823 one finds a thoroughly modern determination of the derivative of the trigonometric functions.

The inequality $1 > \dfrac{\sin \theta}{\theta} > \cos \theta$, for small values of θ, had been implied in the works of Archimedes, and Cauchy indicated that from this inequality it is clear that

$$\lim_{\theta \to 0} \frac{\sin \theta}{\theta} = 1$$ if the measure of the angle is

in radians. Cauchy then formed, for $y = \sin x$, the difference quotient or "Cauchy fraction"

---

tion of the differential of the arcsine from the complex logarithmic relationship appeared also a year earlier in the *Traité du calcul intégral* of L. A. Bougainville (Paris, 1754), pp. 22–24. This work may well have resulted from the influence of Euler's many published papers.

[21] *Cours complet de mathématiques* (5 vols., Paris, 1774), I, xij; III, 36 ff. This work was reprinted at Paris in 1778.

[22] *Traités élémentaires de calcul différentiel et de calcul intégral* (Paris, 1775), p. 478 ff. The original Italian edition appeared at Milan in 1748, the year of Euler's *Introductio*.

[23] *Analyse des infiniment petits, pour l'intelligence des lignes courbes* (new ed. revised by Lefèvre-Gineau, Paris, 1781), pp. 30–31, 34–35.

[24] Augustin Cauchy, *Oeuvres complètes* (25 vols., Paris, 1882–1932), series 2. vol. IV, pp. 14, 22, 24.

$$\frac{\Delta y}{\Delta x} = \frac{\sin (x+i) - \sin x}{i}$$

$$= \frac{\sin \frac{1}{2}i}{\frac{1}{2}i} \cos (x + \tfrac{1}{2}i),$$

from which it was clear that the derivative (i.e., the limit as $i = \Delta x$ tends toward zero) is cos x. The derivative of $y = \arcsin x$ he derived in the now customary manner through the implicit function formula.

From the days of Newton and Leibniz to the time of Bolzano and Cauchy the integral had been defined as an anti-derivative. In this interval the Archimedean summation concept had faded into the background, inasmuch as the inverses of fluxions and differentials are more easily determined than are the limits of sums arising in integration. The fundamental theorem of the calculus meanwhile was loosely justified by the simple geometrical observation that the rate of change of the area under a curve is proportional to the ordinate. It was largely the nineteenth century arithmetization of mathematics that led Cauchy to restore the summation concept of integration and to prove that the integral as so defined is the inverse of the derivative.[25] This proof, coupled with the work in differentiation, completes the modern equivalent of the ancient theorem of Archimedes on the sum of chords. In a sense, then, it marks the close of the historical development of the elementary calculus of the sine. During the first twenty centuries of this history trigonometry had remained essentially geometrical, and hence it played only an incidental role in infinitesimal methods. During the middle of the eighteenth century, however, it was quickly absorbed into the calculus after an arithmetization by which it became associated with the idea of functionality. This is an early illustration of a fact which has become increasingly evident ever since: the basis of trigonometry, of the calculus, and of all modern analysis is found in that central theme of mathematics—the function concept.

[25] Cauchy, *op. cit.*, p. 159 f.

The Trigonometric Functions

By JOHN W. CELL

Professor of Mathematics, North Carolina State College

IN THIS article we shall indicate many different methods by which the sine and cosine functions may be *defined*. (From these the other four functions may be obtained by their usual definitions in terms of the sine and cosine functions, viz., $\cot \theta = \cos \theta / \sin \theta$.) In the course of the discussion we shall consider the trigonometric functions from various points of view and we shall list properties which are not to be found in standard texts on trigonometry but which are found in advanced mathematics. We shall also indicate some general applications which are inherent in these various methods and sources for other applications.

I. DEFINITIONS IN TERMS OF ORDINATE, ABSCISSA, AND DISTANCE

Of all the definitions of the trigonometric functions, the most important are the set in which we define

$$\sin \theta = \frac{\text{ordinate}}{\text{distance}} = \frac{y}{r},$$

$$\cos \theta = \frac{\text{abscissa}}{\text{distance}} = \frac{x}{r}.$$

From these two definitions and the Pythagorean theorem it is easy to prove that $\sin^2 \theta + \cos^2 \theta \equiv 1$. (Students need to learn how to obtain easily and quickly the other two Pythagorean identities from

this identity. Too many students lack confidence in their knowledge of trigonometric identities—a defect which stems from trying to memorize all of them instead of memorizing a few from which they could obtain the others.)

The two basic reduction rules are derivable directly from these ordinate-abscissa-distance definitions: the rules for simplifying

any function (even $\cdot 90° \pm \theta$),
any function (odd $\cdot 90° \pm \theta$).

We may utilize these same definitions to derive

$$\sin (\theta + \phi) = \sin \theta \cos \phi + \cos \theta \sin \phi,$$

$$\cos (\theta + \phi) = \cos \theta \cos \phi - \sin \theta \sin \phi.$$

The student should be afforded experience in starting with these and obtaining identities for $\tan (\theta + \phi)$, $\sin 2\theta$, $\cos 3\phi$, $\sin (A+B) + \sin (A-B)$, etc. (The typical

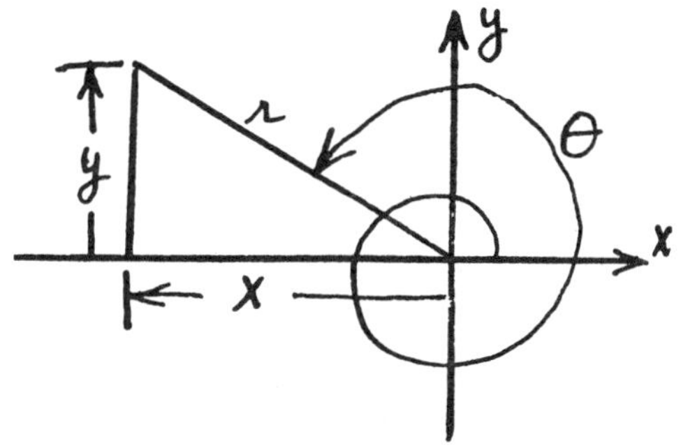

textbook gives too many of such identities in bold-faced type instead of encouraging the student to develop his own ability in performing algebraic manipulations with the basic identities. Notice that all of the fundamental identities that the student *must* memorize can be written in longhand on one side of an ordinary sheet of paper.)

The student of calculus must have ready command of this "algebra of trigonometry"; for instance he should find it easy to simplify $\sqrt{4-9x^2}$ if $3x=2\sin\theta$, to change $\cos^4 7\theta$ to $[(1+\cos 14\theta)/2]^2$ to $\frac{3}{8}+\frac{1}{2}\cos 14\theta+\frac{1}{8}\cos 28\theta$, to change $\cos 7\theta\sin 3\theta$ to $\frac{1}{2}\sin 10\theta-\frac{1}{2}\sin 4\theta$ (here he ought to *start* by writing out the expansions of $\sin(7\theta+3\theta)$ and $\sin(7\theta-3\theta)$). Moreover, applications of this "algebra of trigonometry" are legion in texts in physics and engineering.

Turning from algebraic to graphical trigonometry, we may from these same definitions calculate the values of the trigonometric functions of special angles (0°, 30°, 45°, etc.). We may plot the usual graphs of $y=\sin x$ and $y=\cos x$, utilizing the angle x in radian measure. We may interpret these same equations differently and use $r=\sin\theta$ and $r=\cos\theta$, and draw the corresponding circles in these polar coordinates. Applications of these circle diagrams and of the usual sine and cosine graphs are to be found in any problem involving vibration. For instance, the common home voltage wave is of the form $y=169\sin 120\pi t$ with t in seconds and y in volts. That this voltage is commonly designated as being about 120 volts can be explained by the following graphical method: Suppose that we interpret the equation in the form $r=169\sin\theta$ with $\theta=120\pi t$ radians. In one second the angle θ will go through 120π radians or 60 revolutions or 60 cycles. If, now, we plot the graph in polar coordinates, we obtain a circle with center on the ninety-degree line and radius 169/2. As θ takes on values from 0 to 2π radians, the radius vector traces this same circle twice. The total area traced out in one full revolution is, then, $2(169/2)^2\pi$ or $169^2\pi/2$. Now a circle, with center at the pole and radius such that the area of this new circle is the same as the preceding area, will have a radius given by $\pi r^2=169^2\ \pi/2$, whence $r=169/\sqrt{2}$, or about 120. If, instead of $r=120$, we revert to the rectangular coordinate interpretation for the same equation, we have $y=120$ volts .(Other explanations for this are to be found in texts on electricity.)

II. Definitions in Terms of the Opposite Side, Adjacent Side, and Hypotenuse

If θ is one of the acute angles of a right triangle, then

$$\sin\theta=\frac{\text{opposite}}{\text{hypotenuse}},$$

$$\cos\theta=\frac{\text{adjacent}}{\text{hypotenuse}}.$$

For acute angles these definitions are equivalent to the definitions in terms of the ordinate, abscissa, and distance; if the angle θ is NOT between 0° and 90° the right-triangle definitions do NOT apply.

Every text on trigonometry has applications of these definitions in terms of mensuration problems involving right and oblique triangles. The solution of oblique triangles has been taught traditionally in terms of the familiar four cases and with the laws of sines, cosines, tangents, and half-angle formulas. With the prevalent usage of high-speed computing machines, the rôle of the laws of sines and cosines has become dominant. The chief importance that we teachers can attach to the laws of tangents and half-angles is in their derivations.

III. Definitions as Line Values

That we may define $\sin\theta$ and $\cos\theta$ as line values, or as projections of a rotating vector, is common knowledge. Moreover, such definitions are readily seen to be equivalent to the definitions in terms of ordinate, abscissa, and distance.

This concept finds application in discussions of problems involving vibration (mechanical engineering, aeronautics, electrical engineering).

We teachers find the concept of line values an aid in teaching the student the fundamental identities and the reduction formulas. It is a convenient mode of thought.

IV. THE EULER DEFINITIONS IN TERMS OF $e^{i\theta}$

The sine and cosine functions were defined by Euler (1707–1783) as follows:

$$\sin \theta = (e^{i\theta} - e^{-i\theta})/(2i),$$

$$\cos \theta = (e^{i\theta} + e^{-i\theta})/2,$$

where $e \approx 2.718$ and $i = \sqrt{-1}$. From these definitions one can establish such identities as $\sin^2 \theta + \cos^2 \theta = 1$, $\cos 2\theta = \cos^2 \theta - \sin^2 \theta$; the resulting practice in the laws of exponents is of value to the student who expects to take two years or more of college mathematics.

That these Euler definitions are equivalent to the definitions in terms of the ordinate, abscissa, and distance is not readily apparent. The equivalence is established by the fact that the sine and cosine functions, whether defined by the one method or the other, satisfy basic identities and have properties of continuity and differentiability. In this and the succeeding sections we shall step from trigonometry as such into various sections of advanced mathematics, from which vantage points we can describe other interesting properties of the trigonometric functions.

From these two Euler definitions we can show that

$$e^{i\theta} = \cos \theta + i \sin \theta.$$

We note, for example, that when $\theta = 3\pi$, $e^{i3\pi} = -1$; hence one value of $\log_e (-1)$ is $i3\pi$. More generally,

$$\log_e (-1) = 1n(-1) = i(2n+1)\pi$$

with $n = 0$, ± 1, ± 2, etc. This example illustrates the fact that there are an infinitude of logarithms of any given number. For each positive real number, there is one logarithm that is a real number and an infinitude of logarithms that are of the form: the real logarithm $+ i2n\pi$.

A further consequence may be obtained from the result for $\theta = \pi$: $e^{i\pi} = -1$; and from the knowledge that e is a transcendental number, one can establish by methods of advanced mathematics that π is likewise a transcendental number. From this fact one can show that the famous Greek problem of squaring the circle is unsolvable.

From these same definitions it is relatively easy to compute the sine of a complex angle as, for example, $\sin (3+4i)$—an ability needed by the advanced student in aeronautics and electrical engineering.

V. DEFINITIONS AS SOLUTIONS OF FUNCTIONAL EQUATIONS

Of interest to teachers of trigonometry are the definitions as solutions of functional equations. We may define $\sin x$ and $\cos x$ as those solutions of the two functional equations

$$S(2x) = 2 S(x) C(x),$$

$$C(2x) = 2[C(x)]^2 - 1,$$

which satisfy the conditions that $S(0) = 0$ and $C(0) = 1$, and which satisfy the further condition that $C(\pi/2) = 0$ and $C(x) \neq 0$ for $0 < x < \pi/2$.

There is no special reason for using these particular identities; we could use any two of the usual fundamental identities as the basis of this type of definition. In other words, all of the algebraic portion of trigonometry is derivable from a pair of the identities together with suitable special values for boundary requirements.

Methods of solving such functional equations are to be found in texts in advanced mathematics. Such methods basically amount to changing the problem into one involving differential equations or some other type of equation where the procedure for solution is well known.

This type of definition is used in advanced mathematics to define certain functions.

VI. Definitions in Terms of Infinite Series

The definitions in terms of infinite power series are especially useful for computation and hence for construction of tables of trigonometric functions.

We may begin afresh and define

$$\sin x = x - \frac{x^3}{3!} + \frac{x^5}{7!} - \cdots ,$$

$$\cos x = 1 - \frac{x^2}{2!} + \frac{x^4}{4!} - \cdots ,$$

with the angle x in radian measure. From the rules of infinite series it can be shown that these series converge (and can be used) for *all* values of x. However, they are best suited for computation with small values of x, since the terms will then decrease rapidly in value.

One may use these two infinite series and the rules that govern the algebraic manipulation of infinite series to show that $\sin 2x = 2 \sin x \cos x$, for example, and that these definitions are equivalent to those in terms of the ordinate, abscissa, and distance.

We observe that if x is quite small, then $\sin x \approx x$. A better approximation, clearly, would be $\sin x \approx x - x^3/6$, and so on. Thus, $\sin (0.1 \text{ radian}) \approx 0.1$ and $\cos (0.1) \approx 0.995$.

We may illustrate the use of these series by computing $\sin (1°)$ and $\cos (1°)$ as follows:

$$\sin (1°) = \sin (\pi/180 \text{ rad.})$$

$$= (\pi/180) - \frac{(\pi/180)^3}{3!}$$

$$+ \frac{(\pi/180)^5}{5!} - \cdots$$

$$= 0.0174533 - 0.00000088 + \cdots$$

$$\approx 0.017453,$$

$$\cos (1°) = 1 - (\pi/180)^2/2 + (\pi/180)^4/24$$

$$- \cdots \approx 0.999848.$$

From these results and the knowledge of the values of the sine and cosine of the special angles we may compute, for example, the values of $\sin 31°$ and $\cos 44°$.

VII. Definitions as Solutions of Differential Equations

Teachers of trigonometry will find in this and in the following sections facts which, while of little value in their teaching, will nevertheless be of interest in giving a rounded picture of trigonometry.

We may define $\sin x$ and $\cos x$ as those solutions of the differential equation

$$\frac{d^2 y}{dx^2} + y = 0,$$

which satisfy certain initial or boundary conditions. For $\sin x$ we could require that $y = \sin x = 0$ when $x = 0$, and that $dy/dx = 1$ when $x = 0$. For $y = \cos x$ we could require that $y = 1$ when $x = 0$, and $dy/dx = 0$ when $x = 0$.

We may solve this differential equation for the two solutions that satisfy the two given sets of numerical conditions. If we solve for the two solutions in infinite power-series form, we will obtain the infinite series already used to define these two functions in section VI. Alternately, we may assume $y = ae^{mx}$ to be a solution by which we are led to the solution of the algebraic equation $m^2 + 1 = 0$. If we then take the general solution to be $y = ae^{ix} + be^{-ix}$ and apply the given numerical conditions, we are led to the Euler definitions in terms of the exponential function as given in section IV.

As another method of using differential equations, we could define $y = \sin x$ and $z = \cos x$ as those solutions of the simultaneous differential equations

$$dy/dx = z \text{ and } dz/dx = -y$$

that satisfy the conditions: $y = 0$ and $z = 1$ when $x = 0$.

Differential equations are used as the genesis of many functions of advanced mathematics. For example, the Bessel functions are often defined as solutions of a particular differential equation from which the properties of the Bessel functions are derived.

VIII. Definitions as Definite Integrals

The multifarious facets of trigonometric functions are further exemplified in integral calculus. We define

$$\text{arc sin } y = \int_0^y \frac{dw}{\sqrt{1-w^2}},$$

$$\text{arc tan } y = \int_0^y \frac{dw}{1+w^2}.$$

Since these definitions define these two inverse trigonometric functions only for the range of principal values, we need to adjoin to these definitions certain requirements of continuity or of periodicity in order to obtain the equivalent of the general definitions already given.

Many functions are defined, as regards their genesis, as definite integrals. For example, the gamma function satisfies

$$\Gamma(n) = \int_0^\infty x^{n-1} e^{-x} dx, \quad n > 0.$$

On the other hand, if n is a positive integer, $\Gamma(n+1) = n!$. Articles which treat new functions defined as definite integrals are numerous in the advanced journals of science, engineering, and mathematics.

IX. Definitions as Special Series

If we integrate by parts the two definite integrals treated in the preceding section, we are led to the following series:

$$\text{arc sin } y = \frac{y}{(1-y^2)^{1/2}} - \frac{1}{3} \frac{y^3}{(1-y^2)^{3/2}}$$
$$- \frac{1}{5} \frac{y^5}{(1-y^2)^{5/2}} - \cdots ,$$

$$\text{arc tan } y = \frac{y}{1+y^2} + \frac{2}{3} \frac{y^3}{(1+y^2)^2}$$
$$+ \frac{2}{3} \frac{4}{5} \frac{y^5}{(1+y^2)^3} + \cdots .$$

The first series converges (is usable) for $-0.707 < y < 0.707$, and the second series

converges for all real values of y. These two series may, of course, be used as the genesis of the algebraic portion of trigonometry.

The prime reason for mentioning these peculiar series is that when certain other definite integrals for functions of advanced mathematics are integrated in the same manner, the resulting series are divergent but are also in a special class of series which are known as asymptotic series and may be used, subject to certain rules, for computation purposes.

X. Definitions as Infinite Products

We may define $\sin x$ and $\cos x$ anew by the following infinite products:

$$\sin x = x \prod_{n=1}^\infty \left(1 - \frac{x^2}{\pi^2 n^2}\right)$$
$$= x \left(1 - \frac{x^2}{\pi^2}\right) \left(1 - \frac{x^2}{4\pi^2}\right) \cdots ,$$
$$\cos x = \prod_{n=1}^\infty \left(1 - \frac{4x^2}{\pi^2 (2n-1)^2}\right).$$

These infinite products portray the sine and cosine functions in factored form. These are equivalent definitions for these two trigonometric functions whose proof involves advanced mathematics.

XI. Definitions as Continued Fractions

We could, if we wished, start with the following as the definition for $\tan x$:

$$\tan x = \cfrac{x}{1 - \cfrac{x^2}{3 - \cfrac{x^2}{5 - \cfrac{x^2}{7 - \cdots}}}}$$

This definition is valid for all values of x except, of course, those values of x for which $\tan x$ is not defined.

If we compute the successive values of this fraction for $x = 1$, we obtain the following sequence of values: $1/1 = 1$,

$1/(1-\frac{1}{3})=1.5$, 1.55556, 1.55738, etc. The value of tan (1 radian) is 1.5574 correct to four decimals.

These eleven methods by which we have defined the trigonometric functions are not all of the methods possible. However, these are sufficient to permit us to say that there are many different ways of defining the trigonometric functions. The definitions given in the latter part of this paper could properly be made the basis of an unusual and stimulating summer course in advanced calculus for teachers of high school and elementary college mathematics.

The APPRECIATION of RADIAN MEASURE in ELEMENTARY CALCULUS

By **SALLY IRENE LIPSEY**
and **WOLFE SNOW**

Brooklyn College
City University of New York
New York, New York

ALTHOUGH a substantial number of results in the calculus of trigonometric functions are contingent on expressing the measure of an angle in radians, many students persist in the use of degree measure and seem to feel more comfortable with it. They convert from radians to degrees, at times erroneously, for their computations and diagrams. How can we persuade them of the importance of radian measure? The following suggestions may help.

Remind students of the historical background of the sexagesimal system, noting that the custom of dividing the circle into 360 equal parts probably dates back to the ancient Babylonians and is related to the time required to travel a Babylonian mile (see Eves 1969, p. 31). Since the world discarded the Babylonian mile long ago, there is no logical basis, other than habit, for the use of degree measure. In modern mathematics, radian measure is far more useful.

Provide review exercises in calculating the length, s, of the arc of a circle subtended by a central angle. If the measure of the angle is θ in degrees, then $s = \pi r \theta / 180$; if α radians are used, $s = r\alpha$. Similarly, have students find the area of a circular sector using both degree measure and radian measure, emphasizing that the computations in the first case involve multiplication by the additional factor $\pi/180$.

Use enough details to illustrate convincingly that the value of $\lim_{x \to 0} (\sin x)/x$ depends on the units of measure. First guide the students in setting up a table to compare various values of $(\sin \theta)/\theta$ with

$(\sin \alpha)/\alpha$ where θ is the measure of a given angle in degrees and α is the measure of the same angle in radians (see fig. 1). The table shows that as θ and α approach 0, $(\sin \theta)/\theta$ and $(\sin \alpha)/\alpha$ seem to be approaching 0.01745 and 1, respectively. Then discuss the standard proof that $\lim_{\alpha \to 0} (\sin \alpha)/\alpha = 1$, which is based on the inequality $\sin \alpha < \alpha < \tan \alpha$ for α in the first quadrant and expressed in radians. Immediately thereafter, the students can find $\lim_{\theta \to 0} (\sin \theta)/\theta$ by converting from degree measure to radian measure. Since θ and α are both measures of the same angle, $\theta = 180\alpha/\pi$ and $\sin \theta = \sin \alpha$. Thus students will see that $\lim_{\theta \to 0} (\sin \theta)/\theta = \lim_{\alpha \to 0}(\sin \alpha)/(180\alpha/\pi) = (\pi/180) \lim_{\alpha \to 0}(\sin \alpha)/\alpha = \pi/180$, which is approximately 0.01745.

Emphasize the role of $\lim_{x \to 0} (\sin x)/x$ in deducing the formula for the derivative of $\sin x$. Discuss the fact that if x represents radian measure, $\dfrac{d}{dx} \sin x = \cos x$, but if x represents degree measure, $\dfrac{d}{dx} \sin x = (\pi/180) \cos x$. Point out that the maximum value of the instantaneous rate of change of the sine of an angle is 1 with respect to its radian measure but $\pi/180$ with respect to its degree measure.

θ	α $(\pi\theta/180)$	Sin θ (sin α)	Sin θ/θ	Sin α/α
10°	0.17453	0.17365	0.01736	0.99496
5°	0.08727	0.08716	0.01743	0.99874
2°	0.03491	0.03490	0.01745	0.99971
1°	0.01745	0.01745	0.01745	1.00000

Fig. 1

Interpret the above results graphically. Have the students sketch the graphs of

$y = \sin x$ on the same axes, assuming first that x represents the number of degrees and then that x represents the number of radians (see fig. 2). This activity reviews the fact that x radians and x degrees produce different values for y. The combined drawings provide a geometric interpretation of the two different formulas for the derivative, making it vividly clear that one graph rises much more rapidly than the other and that the slope of the first never reaches as high a value as the slope of the second.

In general, comparisons of results from computations, sketches, and proofs involving degrees with those involving radians serve to convince students that it is worth their while to learn to use radians. In this paper we have reported some methods that have been successful in clarifying the distinction between degree measure and radian measure and in motivating students to use the latter.

REFERENCE

Eves, Howard. *An Introduction to the History of Mathematics*. 3d ed. New York: Holt, Rinehart & Winston, 1969.

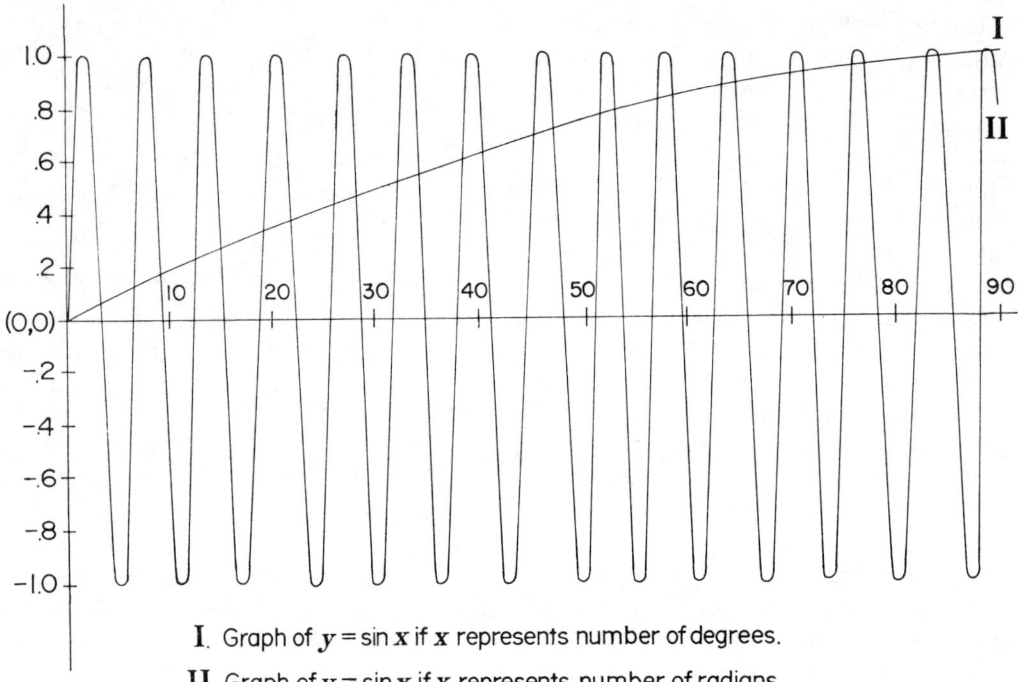

I. Graph of $y = \sin x$ if x represents number of degrees.

II. Graph of $y = \sin x$ if x represents number of radians.

Fig. 2

Bibliography: *Trigonometric Functions*

Tierney, John A. Trigonometric Functions of Real Numbers. 50 (January 1957): 38–39.
Remarks in favor of teaching trigonometric functions as functions of real numbers rather than as functions of angles.

Yates, Robert C. The Trigonometric Functions. 51 (March 1958): 191–93.
Remarks on an approach to defining the trigonometric functions based on expansions of power series.

———, The Euler Identity. 51 (April 1958): 266.
A brief proof of the Euler identity without using infinite series.

Coleman, A. J. Letter to the editor. 51 (October 1958): 474.
A brief proof of the Euler identity without using infinite series.

Hamming, R. W. A Note on the Teaching of Trigonometry. 64 (March 1971): 249–51.
Remarks on an approach to the teaching of trigonometric identities based on the Euler identity.

Hyperbolic Functions

By Aaron Bakst

135-12 77th Avenue, Flushing 67, New York

The development of *hyperbolic functions* in the traditional trigonometry courses (if this is ever reached during a one-semester instruction) is usually confined to purely algebraic methods. However effective the latter procedures may be, it is doubtful that a student realizes the import of the properties of hyperbolic functions. The student is never offered the opportunity to realize the fact that, essentially, the properties of hyperbolic functions are analogous to the properties of circular functions. It is possible, however, to develop the properties of hyperbolic functions in a manner which is analogous to the processes which are employed in the development of circular functions. Thus, it is proposed to examine and to develop hyperbolic functions by means of a geometric approach.

We shall consider the equilateral hyperbola

$$xy = a \qquad (1)$$

whose properties will be examined presently. We shall formulate these properties in terms of certain propositions.

Proposition 1. If the lengths of the straight line segments on the coordinate axis OX which are bounded by the origin

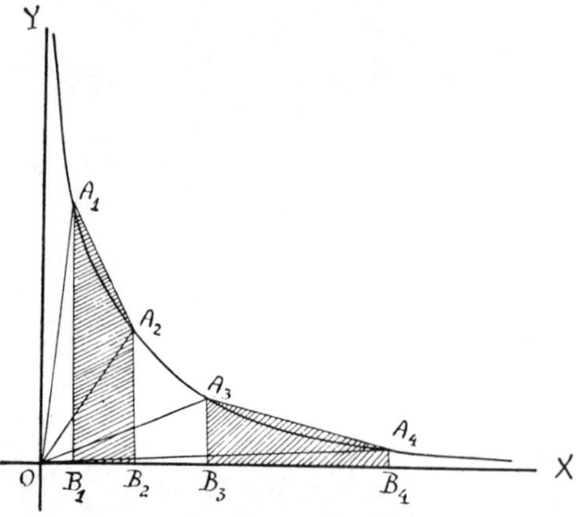

Fig. 1.

and the projections B_1, B_2, B_3, and B_4 of the points A_1, A_2, A_3, and A_4 respectively on the equilateral hyperbola $xy = a$ are proportional, then the areas of the rectangular trapezoids $A_1A_2B_2B_1$ and $A_3A_4B_4B_3$ are equal.

The areas of the trapezoids $A_1A_2B_2B_1$ and $A_3A_4B_4B_3$ (Fig. 1) are

$$S_1 = \tfrac{1}{2}(x_2 - x_1)(y_1 + y_2) \quad \text{and}$$
$$S_2 = \tfrac{1}{2}(x_4 - x_3)(y_3 + y_4) \qquad (2)$$

respectively.

Replacing the y's in the equation (2) by their corresponding expressions in terms of the x's with their appropriate subscripts (see equation (1)) and performing the algebraic simplifications, we obtain

$$S_1 = a \left(\frac{x_2}{x_1} - \frac{x_1}{x_2} \right) \quad \text{and}$$
$$S_2 = a \left(\frac{x_4}{x_3} - \frac{x_3}{x_4} \right). \qquad (3)$$

Since

$$x_1 : x_2 = x_3 : x_4, \quad S_1 = S_2.$$

The converse of Proposition 1 may be established by equating the relations (3). Let us denote $x_2 : x_1$ by m and $x_4 : x_3$ by n. We then have

$$m - \frac{1}{m} = n - \frac{1}{n}.$$

We then obtain $m^2 n - n = n^2 m - m$ or $m^2 n + m = n^2 m + n$. From this we obtain $m = n$ or $x_1 : x_2 = x_3 : x_4$.

Proposition 2. If the lengths of the straight line segments on the coordinate axis OX which are bounded by the origin and the projections B_1, B_2, B_3, and B_4 of the points A_1, A_2, A_3, and A_4 respectively on the equilateral hyperbola $xy = a$ are proportional, then the areas of the curvilinear trapezoids $A_1A_2B_2B_1$ and $A_3A_4B_4B_3$ are equal.

Divide each of the straight line segments A_1A_2 and A_3A_4 (Fig. 1) into n equal parts (n an integer) and mark off on the hyperbola the corresponding ordinates of the abscissas of the division points thus obtained. Join these points on the hyperbola with their corresponding abscissas by straight line segments. Furthermore, join the points on the hyperbola thus obtained by straight line segments. We thus obtain two polygonal areas, each of them consisting of n trapezoids. The two polygonal areas are equal according to Proposition 1. This may be noted from the fact that the first small trapezoid in $A_1A_2B_2B_1$ has an area which is equal to the area of the first trapezoid in $A_3A_4B_4B_3$. The same may be observed for the pair of the second trapezoids in the two polygons, and so on. Allowing n to approach infinity and considering the limiting case, we may obtain the limits of the two polygonal areas $A_1A_2B_2B_1$ and $A_3A_4B_4B_3$. The limits of these two polygonal areas are equal.[1]

It may be noted that the converse of Proposition 2 is also true. The proof of this converse follows the procedures used in the proofs of the above two propositions.

PROPOSITION 3. If the lengths of the straight line segments on the coordinate axis OX which are bounded by the origin and projections B_1, B_2, B_3, and B_4 of the points A_1, A_2, A_3, and A_4 respectively on the equilateral hyperbola $xy = a$ are proportional, then the areas of the sectors OA_1A_2 and OA_3A_4 are equal (Fig. 1).

Let us denote the area of the curvilinear trapezoid by $(A_1A_2B_2B_1)$. Then Area of sector $OA_1A_2 =$ Area of $(A_1A_2B_2B_1) +$ Area of $\triangle OA_1B_1 -$ Area of $\triangle OA_2B_2$.

By virtue of the equation of the equilateral hyperbola $xy = a$,

$$\text{Area of } \triangle OA_1B_1 = \text{Area of } \triangle OA_2B_2.$$

Thus,

Area of sector $OA_1A_2 =$ Area of $(A_1A_2B_2B_1)$

Similarly,

Area of sector $OA_3A_4 =$ Area of $(A_3A_4B_4B_3)$.

Then, by virtue of Proposition 2,

Area of sector OA_1A_2

$$= \text{Area of sector } OA_3A_4.$$

It should be noted that the converse of this proposition is also true. The proof of the converse may be obtained by establishing the fact that the areas of the triangles OA_1B_1 and OA_2B_2 are equal. From this will follow that the areas of the curvilinear trapezoids $A_1A_2B_2B_1$ and $A_3A_4B_4B_3$ are equal. Then this converse follows from the converse of Proposition 2.

PROPOSITION 4. If the point A is the midpoint of the chord A_1A_2 of the equilateral hyperbola $xy = a$, then the straight lines which are drawn through the points A_1 and A_2 parallel to the asymptotes of the equilateral hyperbola intersect on the straight line OA (Fig. 2).

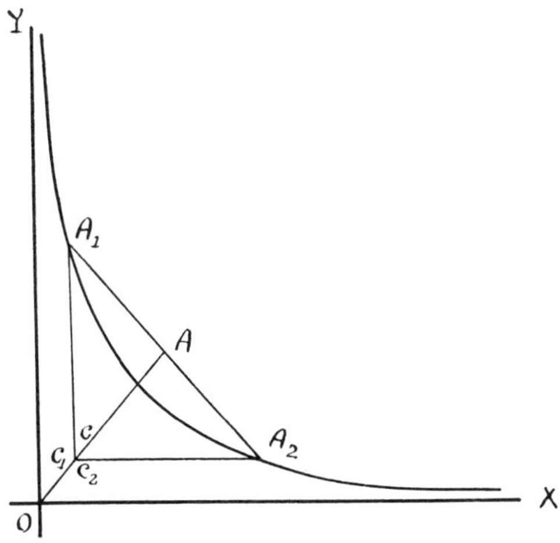

FIG. 2

Since A_1C_1 is parallel to OY,

$$\frac{OA}{OC_1} = \frac{x_1 + x_2}{2x_1}.$$

Since A_2C_2 is parallel to OX,

$$\frac{OA}{OC_2} = \frac{y_1 + y_2}{2y}.$$

---

[1] It should be understood that this is only an intuitive approach to the proof of this proposition. A rigorous proof of this proposition may be obtained by means of integral calculus.

The right members of these two equations are equal by virtue of the equation of the equilateral hyperbola $xy = a$. Hence

$$OA : OC_1 = OA : OC_2,$$

and the points C_1 and C_2 are coincident.

The converse of this proposition is also true.

PROPOSITION 5. If the point A is the midpoint of the chord A_1A_2 of the equilateral hyperbola $xy = a$, then the straight lines OA and A_1A_2 make equal angles with the asymptotes of the equilateral hyperbola (Fig. 3).

with its chord (OA and A_1A_2 respectively) depends on the direction of the chord A_1A_2. Hereafter we shall refer to the directions of OA and A_1A_2 as *conjugate directions*.

PROPOSITION 6. If A is the midpoint of the chord A_1A_2 of the equilateral hyperbola $xy = a$, then the straight line OA intersects the hyperbola in the point A_0 whose coordinates (x_0, y_0) are the mean proportionals of the respective coordinates of the points A_1 and A_2 (Fig. 3).

The equation of the equilateral hyperbola leads to

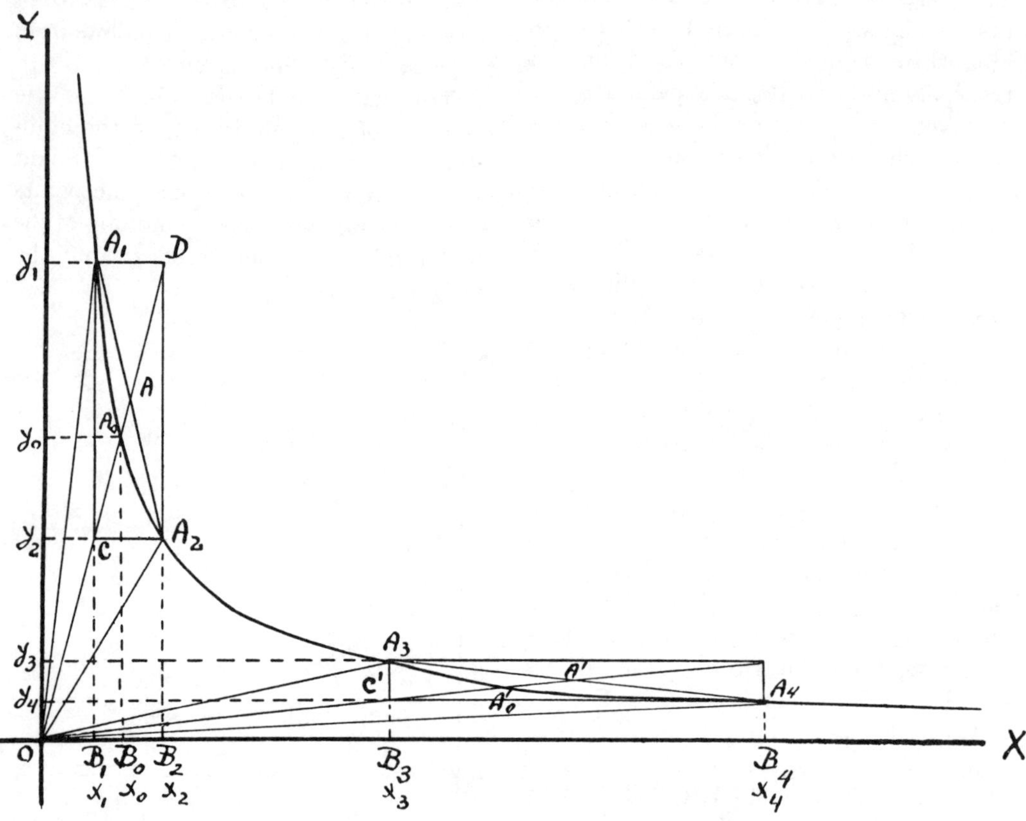

FIG. 3

According to Proposition 4 we have a right triangle A_1CA_2 in which the straight line AC is the median. In other words, the triangles A_1CA and ACA_2 are both isosceles. Then $\angle AA_1C = \angle ACA_1$ and $\angle ACA_2 = \angle AA_2C$.

Proposition 5 is very important because from this theorem it follows that the direction of the diameter which is conjugate

$$x_0y_0 = x_1y_1 = x_2y_2.$$

From the similarity of the triangles ODB_2 and OA_0B_0 we have

$$x_0 : y_0 = x_2 : y_1.$$

Multiplying this equation and the equation $x_0y_0 = x_1y_1$ term by term, we obtain

$$x_0^2 = x_1x_2.$$

Multiplying term by term the equations $x_0 y_0 = x_2 y_2$ and $y_0 : x_0 = y_1 : x_2$ we obtain

$$y_0{}^2 = y_1 y_2.$$

As a corollary, we find that the straight line OA_0 bisects the area of the sector $OA_1 A_2$. This follows from the results of Proposition 3 and from the relation

$$x_1 : x_0 = x_0 : x_2.$$

PROPOSITION 7. If Area of sector $OA_1 A_2$ = Area of sector $OA_3 A_4$ (Fig. 3), and A and A' are the midpoints of the chords $A_1 A_2$ and $A_3 A_4$ of the equilateral hyperbola $xy = a$, then we have the proportion

$$OA_0 : OA : A_1 A_2 = OA_0' : OA' : A_3 A_4.$$

Since the areas of the sectors are equal, we have, by virtue of the converse of Proposition 3,

$$x_1 : x_2 = x_3 : x_4.$$

According to Proposition 4

$$\frac{OA}{OC} = \frac{x_1 + x_2}{2x_1} \quad \text{and} \quad \frac{OA'}{OC'} = \frac{x_3 + x_4}{2x_3}.$$

Then

$$\frac{OA}{OC} = \frac{OA'}{OC'} \quad \text{or} \quad \frac{OA}{OA'} = \frac{OC}{OC'}. \quad (4)$$

Furthermore,

$$\frac{OC}{OD} = \frac{OC'}{OD'}. \quad (5)$$

According to Proposition 6,

$$OA_0{}^2 = OA \cdot OC$$

and

$$OA_0'{}^2 = OA' \cdot OC'.$$

From this it follows that

$$\frac{OA_0}{OA_0'} = \frac{OA}{OA'} = \frac{OC}{OC'}. \quad (6)$$

Since

$$\frac{OD}{OC} = \frac{OC + A_1 A_2}{OC} = 1 + \frac{A_1 A_2}{OC}$$

and

$$\frac{OD'}{OC'} = \frac{OC' + A_3 A_4}{OC'} = 1 + \frac{A_3 A_4}{OC'}.$$

Then, by virtue of (5),

$$\frac{A_1 A_2}{A_3 A_4} = \frac{OC}{OC'}. \quad (7)$$

The proportions (4), (6), and (7), when combined, establish the proof of Proposition 7, that is,

$$OA_0 : OA : A_1 A_2 = OA_0' : OA' : A_3 A_4.$$

From Proposition 7 we obtain another result. The proportions

$$\frac{AA_1}{OA_0} = \frac{A'A_3}{OA_0'} \quad \text{and} \quad \frac{OA}{OA_0} = \frac{OA'}{OA_0'}$$

are functions of the areas of the sectors $OA_1 A_2$ and $OA_3 A_4$ (these areas being equal) for any given equilateral hyperbola. Let us denote the areas of these sectors by T. We may then denote these functional relations as follows:

$$\frac{AA_1}{OA_0} = \sinh T \quad \text{and} \quad \frac{OA}{OA_0} = \cosh T.$$

The construction of the hyperbolic sine and hyperbolic cosine are carried out as follows. According to the corollary of Proposition 6, the straight line OA_0 bisects the area of the sector $OA_1 A_2$. We construct the sector $OA_1 A_0 = 1/2\ T$. Through the point A_1 we draw a straight line $A_1 A$ which should be *conjugate* with the straight line OA_0.[2]

According to Proposition 6 (Fig. 3), $x_0{}^2 = x_1 x_2$.

Then,

---

[2] The definition of the hyperbolic sine and hyperbolic cosine given above is analogous in many respects with the definition and development given by E. W. Hobson in *A Treatise on Plane Trigonometry*, Cambridge: at the University Press, 1921, pp. 329–330. However, the departure here from Hobson's treatment (he draws an analogy with the properties of the circle and the area of a circular sector) consists in the consistent employment of the properties of the hyperbola.

$$\frac{OC}{OA_0}=\frac{x_1}{x_0} \quad \text{and} \quad \frac{OA_0}{OD}=\frac{x_0}{x_2},$$

or,

$$\frac{OC}{OA_0}=\frac{OA_0}{OD}.$$

Then,

$$OA_0{}^2=OC\cdot OD=(OA-AC)(OA+AD)$$
$$=(OA-AD)(OA+AD)$$
$$=OA^2-AD^2,$$
$$OA_0{}^2=OA^2+AA_1{}^2.$$

Hence,

$$\frac{OA^2}{OA_0{}^2}-\frac{AA_1{}^2}{OA_0{}^2}=1,$$

or,

$$\cosh^2 T-\sinh^2 T=1.$$

The derivation of the expression for the hyperbolic functions of the sum of two angles is analogous to the derivation of the expressions for circular functions. We will demonstrate the complete analogy by using two diagrams with identical letterings (Figs. 4a and 4b).

Let Area of sector $OAB=1/2\ a$ and Area of sector $OBC=1/2\ b$ (Fig. 4b). Then, Area of sector $OAC=1/2(a+b)$.

In order to construct the hyperbolic functions we proceed as follows. Draw the straight line BH making the same angles with the asymptotes of the hyperbola as the straight line OA. In other words, BH and OA will have conjugate directions. Then draw CG parallel to BH. In a similar manner, draw CE so that it will be conjugate with OB. Then,

$$\sinh a=\frac{BH}{OA}, \quad \sinh b=\frac{CE}{OB}, \quad \text{and}$$
$$\sinh (a+b)=\frac{CG}{OA}.$$
$$\cosh a=\frac{OH}{OA}, \quad \cosh b=\frac{OE}{OB}, \quad \text{and}$$
$$\cosh (a+b)=\frac{OG}{OA}.$$

The expressions for the circular functions may be obtained from Figure 4a.

Draw in Figures 4a and 4b straight lines EF, BH, and CG parallel to each other. Also, draw ED parallel to OA. Then,

$$\sinh (a+b)=\frac{CG}{OA}=\frac{CD}{OA}+\frac{DG}{OA}=\frac{EF}{OA}+\frac{DC}{OA}.$$

From the similarity of the triangles OEF and OBH we have:

$$EF:BH=OE:OB.$$

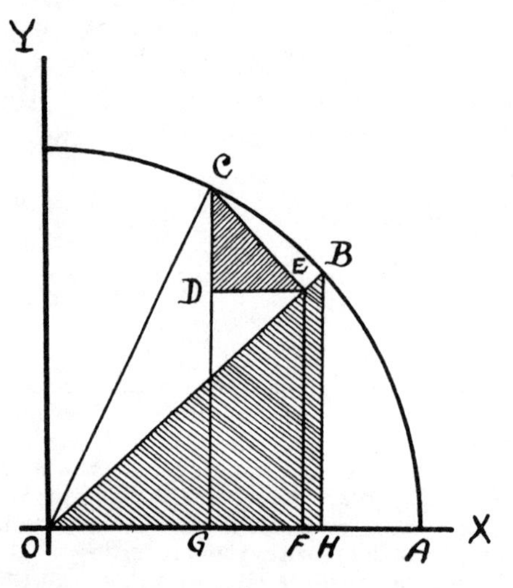

FIG. 4

Then,

$$\frac{EF}{OA}=\frac{BH}{OA}\cdot\frac{OE}{OB}=\sinh a\cdot\cosh b.$$

From the similarity of the triangles DEC and OBH we have:

$$DC:OH=EC:OB.$$

Then,

$$\frac{DC}{OA}=\frac{OH}{OA}\cdot\frac{EC}{OB}=\cosh a\cdot\sinh b.$$

Finally,

$$\sinh (a+b)=\sinh a\cdot\cosh b+\cosh a\cdot\sinh b.$$

Similarly,

$$\cosh (a+b)=\frac{OG}{OA}=\frac{OF}{OA}+\frac{FG}{OA}.$$

Using the similarity properties of the above two pairs of triangles, we have:

$$\frac{OF}{OA}=\frac{OH}{OA}\cdot\frac{OE}{OB}=\cosh a\cdot\cosh b,$$

and

$$\frac{FG}{OA}=\frac{ED}{OA}=\frac{BH}{OA}\cdot\frac{EC}{OB}=\sinh a\cdot\sinh b.$$

Finally,

$$\cosh (a+b)=\cosh a\cdot\cosh b+\sinh a\cdot\sinh b.$$

The process of the derivation of the expressions for hyperbolic functions employs the property of *conjugate directions* with respect to a hyperbola, while the process of the derivation of the expressions for circular functions employs the property of *perpendicularity*. Thus, *perpendicularity* represents *conjugate directions* with respect to a circle.

The function

$$f(a)=\cosh a+\sinh a$$

satisfies the functional equation

$$f(a)\cdot f(b)=f(a+b).$$

We have:

$$f(a)\cdot f(b)$$
$$=(\cosh a+\sinh a)\cdot(\cosh b+\sinh b)$$
$$=\cosh (a+b)+\sinh (a+b).$$

Thus,

$$f(a)\cdot f(b)=f(a+b).$$

The functional equation

$$f(a)\cdot f(b)=f(a+b)$$

leads to the derivation of the expressions for hyperbolic functions

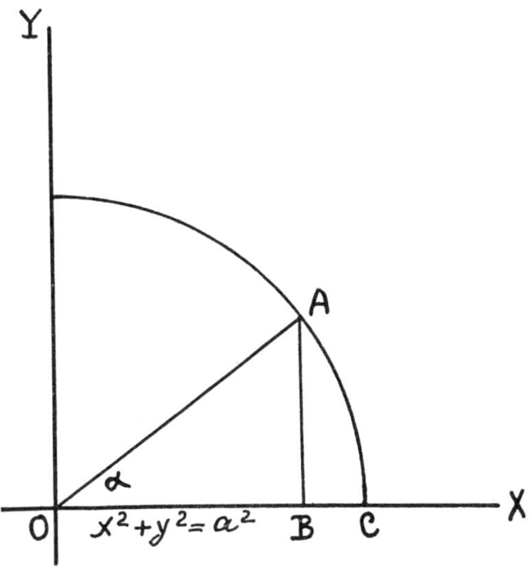

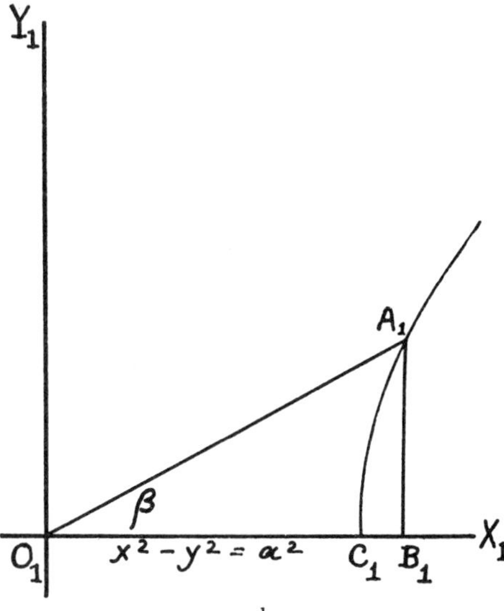

FIG. 5

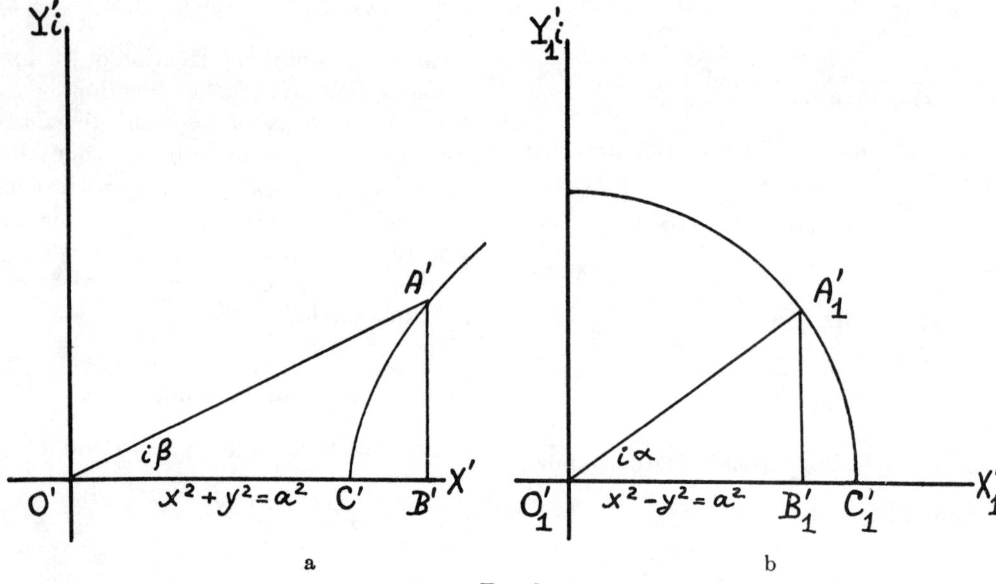

Fig 6

$\sinh a = \frac{1}{2}(e^a - e^{-a})$ and $\cosh a = \frac{1}{2}(e^a + e^{-a})$, where e is the base of the natural logarithms.

The structural relationship between hyperbolic functions and circular functions manifests a striking similarity if these two types of functions are considered in the real plane and in the complex plane.

In the real plane the coordinates of a point in the plane are represented by real numbers. In the complex plane, the abscissas are represented by real numbers, while the ordinates are represented by imaginary numbers. Thus, in the complex plane the coordinate axis OY is known as the *imaginary axis*.

In the real plane, the equation of a circle (with the center at the origin) is $x^2 + y^2 = a^2$, and the equation of a hyperbola (associated with the same circle) is $x^2 - y^2 = a^2$. The graphs of these two curves are represented in Figures 5a and 5b.

The equation of the circle in the complex plane is $x^2 - y^2 = a^2$, while the equation of the hyperbola in the complex plane is $x^2 + y^2 = a^2$. The graphs of these two curves are represented in Figures 6a and 6b.

For the sake of simplicity we shall assume that $a = 1$.

We have then the following expressions:

$$\sin \alpha = AB, \quad \cos \alpha = OB \quad \text{(Fig. 5a)}$$

$$\sin i\beta = A'B', \quad \cos i\beta = O'B' \quad \text{(Fig. 6a)}$$

$$\sinh \beta = A_1B_1, \quad \cosh \beta = O_1B_1 \quad \text{(Fig. 5b)}$$

$$\sinh i\alpha = A_1'B_1',$$

$$\cosh i\alpha = O_1'B_1' \quad \text{(Fig. 6b)}$$

$A_1'B_1' = i \cdot AB$. Therefore, $\sinh i\alpha = i \sin \alpha$.

$O_1'B_1' = OB$. Therefore, $\cosh i\alpha = \cos \alpha$.

$A'B' = i \cdot A_1B_1$. Therefore, $\sin i\beta = i \sinh \beta$.

$O'B' = O_1B_1$. Therefore, $\cos i\beta = \cosh \beta$.

The logarithm and its inverse

ROBERT C. YATES, *College of William and Mary, Williamsburg, Virginia.*
An approach to any topic that is somewhat "out of the ordinary"
is a good thing. Secondary teachers may use this article to refresh
their knowledge of calculus and gain new insight into ideas already familiar.

IT SEEMS IMPORTANT to define the elementary transcendental functions by means of power series before completing a first course in calculus. It is customarily made clear that these series are a real necessity, for example, in calculating function values for entry into tables. Perhaps it is not customary to display the fact that all properties of these functions may be derived directly from their series representations. This is the purpose of this paper. To sharpen the point, we shall discard prior knowledge, pretend ignorance of the function, and accept the series as its only characterization. The discussion here is limited to the logarithm and the exponential, which will parade under the labels L and E, respectively.

We must assume that the term-by-term differentiation of a series defining a function results in a new series which defines the derivative of the function within the same interval of convergence.

1. *The logarithm function.* The function L is defined by the series:

$$L(x) = (x-1) - \frac{(x-1)^2}{2} + \frac{(x-1)^3}{3} - \cdots,$$
$$\text{in } 0 < x \leq 2 \quad (1)$$

wherein we note that $L(1) = 0$ and that $L(x)$ is negative if $x < 1$, positive if $x > 1$. Now

$$L'(x) = 1 - (x-1) + (x-1)^2 - \cdots,$$
$$0 < x < 2$$

is recognizable as the *geometric series* with ratio $(1-x)$. Its sum is $1/x$. Thus

$$L'(x) = \frac{1}{x}$$

or in differential form:

$$dL(x) = \frac{dx}{x}. \qquad (2)$$

2. *Properties of* L. We use the basic form (2) and positive arguments in the following:

(a) Since

$$dL(ax) = \frac{d(ax)}{ax} = \frac{dx}{x} = dL(x)$$

then

$$d[L(ax) - L(x)] = 0,$$

or

$$L(ax) - L(x) = \text{constant}.$$

But for $x = 1$, $L(1) = 0$ and thus the constant has value $L(a)$. Accordingly,

$$L(ax) = L(x) + L(a). \qquad (3)$$

Here, terms of the right member are valid for $0 < x \leq 2$ and $0 < a \leq 2$. Thus the left member is validated for $0 < ax \leq 4$. We are by such means permitted to extend the original interval in step-by-step fashion.

A corollary is established by setting $x = 1/a$ in (3). We may write

$$L(1) = L\left(\frac{1}{a}\right) + L(a)$$

or

$$L\left(\frac{1}{a}\right) = -L(a). \qquad (4)$$

(b) Setting $x = 1/b$ in (3) there follows

$$L\left(\frac{a}{b}\right) = L(a) + L\left(\frac{1}{b}\right)$$

or with (4):

$$\boxed{L\left(\frac{a}{b}\right) = L(a) - L(b)} \qquad (5)$$

(c) Writing (2) in the form $dL(u) = du/u$, we put $u = x^n$ where n is a rational number. Then

$$dL(x^n) = \frac{dx^n}{x^n} = \frac{nx^{n-1}dx}{x^n} = n \cdot \frac{dx}{x}$$

$$= n \cdot dL(x)$$

or

$$d\left[L(x^n) - n \cdot L(x)\right] = 0.$$

Thus

$$L(x^n) - n \cdot L(x) = \text{constant}.$$

Putting $x = 1$, the constant is evaluated as zero and accordingly

$$\boxed{L(x^n) = n \cdot L(x)} \qquad (6)$$

3. *The exponential function.* We now define a new function E by:

$$E(x) = 1 + x + \frac{x^2}{2!} + \frac{x^3}{3!} + \cdots ,$$

$$-\infty < x < +\infty \qquad (7)$$

and note that $E(0) = 1$ and that $E(x) > 0$ for $x \geq 0$. In addition

$$E'(x) = 1 + x + \frac{x^2}{2!} + \frac{x^3}{3!} + \cdots = E(x)$$

or

$$\boxed{dE(x) = E(x)dx} \qquad (8)$$

Thus

$$dx = \frac{dE(x)}{E(x)} = dL[E(x)]$$

or

$$d\{x - L[E(x)]\} = 0.$$

Then

$$x - L[E(x)] = \text{constant}.$$

But for $x = 0$, $E(0) = 1$ and $L(1) = 0$, and thus this constant is zero. We have then

$$x = L[E(x)],$$

an inverse relation; i.e., L and E are functions such that

$$L(E) = E(L) = I, \text{ the identity.}$$

4. *Properties of* E.

(a) If we let $y = L(x)$ and $k = L(a)$ so that $x = E(y)$ and $a = E(k)$, then

$$y + k = L(x) + L(a) = L(xa).$$

Taking the inverse of this expression, we find

$$\boxed{E(y+k) = xa = E(y) \cdot E(k)} \qquad (9)$$

(b) Setting $z = E(ax)$ so that $L(z) = ax$, then for rational a,

$$x = \frac{1}{a} \cdot L(z) = L(z^{1/a}) \text{ by (6).}$$

The inverse is

$$E(x) = z^{1/a} = [E(ax)]^{1/a}$$

or

$$\boxed{[E(x)]^a = E(ax)} \qquad (10)$$

Note that for $a = -1$, equation (10) yields

$$E(-x) = \frac{1}{E(x)}$$

which, in view of the remark following equation (7), discloses the fact that $E(x)$ is positive for all values of x. Furthermore, as $x \to \infty$, $E(x) \to \infty$ and $E(-x) \to 0$.

For the sake of familiarity, we list the last three forms in their usual garb:

$$dE(x) = E(x)dx \qquad de^x = e^x dx \qquad (8)$$

$$E(y+k) = E(y) \cdot E(k) \qquad e^{y+k} = e^y \cdot e^k \qquad (9)$$

$$[E(x)]^a = E(ax) \qquad (e^x)^a = e^{ax}. \qquad (10)$$

Logarithmic and exponential functions —a direct approach[*]

C. L. SEEBECK, JR., *University of Alabama, and* P. M. HUMMEL,
University of Alabama, University, Alabama.

*A direct approach to logarithms rather than
the indirect approach as an inverse of an exponential function
has many advantages for teaching secondary school students.*

THE USUAL DEVELOPMENT of logarithms given in high school courses has several shortcomings. The logarithmic function is defined as the inverse of an exponential function. Since the student is familiar only with the integral, or at best rational, exponents, this approach does not help him see the uses of logarithms for computational purposes. Neither does it add much to his understanding of functions. Alternate definitions in terms of convergent series, or in terms of integrals, are beyond the mathematical background of most high school students. The direct approach, as a function having certain postulated properties, can easily be followed by elementary high school students. This approach does not use limits or the concepts of continuity. It also serves as a basis for later extensions and gives the student an excellent example of the function concept early in the curriculum.

In a first course the development should be limited to the common logarithm and the theorems necessary for computation. Later on, the theory can be extended to the general base, and the change-in-base procedure established. We shall develop the theory for the general base.

Let x be a member of the set of all positive real numbers and let y be a member of the set of all real numbers. Let log be the symbol for a function which establishes a correspondence between x and y such that for each x there is one and only one y. If y is the number that corresponds to x, we write $y = \log(x)$. The logarithmic function is by definition the function that has the following properties:

(1) $\log(ab) = \log(a) + \log(b)$.
(2) There is a positive number c, called the base, for which $\log(c) = 1$.
(3) If $a > 1$, then $\log(a) > 0$.

When several bases are being used simultaneously, or if we wish to make clear the base being used, we will change notation slightly and write $y = \log_c(x)$, read, "y equals the log of x to the base c." But when only one base is being used and it is clear what it is, the subscript is unnecessary and will not be used.

Postulate (3) is stronger than necessary, and later we will change it and note the changes in the theory. In an elementary course, however, postulate (3) as stated above is the recommended one.

On the basis of the given postulates we next proceed to the proofs of the basic theorems.

[*] The authors wish to acknowledge the many contributions of the Mathematics Staff at the University of Alabama to the development given here.

Theorem 1: $\log(1) = 0$.

Proof: Since $a = a \cdot 1$, set $b = 1$ in (1) getting
$$\log(a) = \log(a) + \log(1)$$
so that
$$0 = \log(1).$$

Theorem 2: $\log(1/a) = -\log(a)$.

Proof: Since $1 = a(1/a)$, set $b = 1/a$ in (1) and get
$$\log(1) = \log(a) + \log(1/a).$$

But $\log(1) = 0$ by theorem 1, so $\log(1/a) = -\log(a)$.

Theorem 3: $\log(a/b) = \log(a) - \log(b)$.

Proof:
$$\log(a/b) = \log\left(a \cdot \frac{1}{b}\right) = \log(a) + \log(1/b)$$
$$= \log(a) - \log(b).$$

Theorem 4: If $a > b$, then $\log(a) > \log(b)$.

Proof: If $a > b$, then $a/b > 1$ and by posstulate (3) $\log(a/b) > 0$.

Hence, $\log(a/b) = \log(a) - \log(b) > 0$, and $\log(a) > \log(b)$.

This last theorem is very important and yields numerous interesting and valuable results. It proves essentially that the logarithmic function is strictly monotonic.

Corollary 4.1. $\log(a) = \log(b)$, if and only if $a = b$.

For if a and b are different, one of them is larger than the other and the theorem applies.

Corollary 4.2. If $\log(b) > \log(a)$, then $b > a$.

This follows since $b < a$ contradicts Theorem 4, and the condition $b = a$ requires $\log(a) = \log(b)$.

Corollary 4.3. The base c is greater than 1.

Corollary 4.4. If $a < 1$, then $\log(a) < 0$.

Thus, the logarithms of numbers between 0 and 1 are negative.

Theorem 5: If n is any rational number, then $\log(a^n) = n \log(a)$.

Proof: Case 1. First, let n be a positive integer. Then
$$\log(a^n) = \log(a \cdot a \cdots a) \text{ with } n \text{ factors}$$
$$= \log(a) + \log(a) + \cdots + \log(a)$$
$$\text{to } n \text{ terms}$$
$$= n \log(a).$$

Case 2. Let $n = p/q$ where p and q are positive integers.

Set $x = a^n = a^{p/q}$. Then $x^q = a^p$ and $\log(x^q) = \log(a^p)$.

Since the theorem holds for case 1, we get
$$q \log(x) = p \log(a), \text{ and } \log(x) = \frac{p}{q} \log(a).$$
Hence,
$$\log(a^n) = n \log(a).$$

Case 3. Finally, let $n = -r$, where r is a positive rational number. Then
$$\log(a^n) = \log(1/a^r) = -\log(a^r)$$
$$= -r \log(a) = n \log(a).$$

Corollary 5.1. If r is any rational number, then $\log(c^r) = r$.

To show that properties (1), (2), and (3) determine a unique function, we need the following:

Auxiliary theorem. If a and b are any two real numbers with $a < b$, then there is a rational number r such that $a < r < b$.

Proof: Since $a < b$, $b - a > 0$, and for a sufficiently large positive integer N, $b - a > 1/10^N$. Now consider the set of rational numbers $0, 1/10^N, 2/10^N, 3/10^N, \cdots$, together with the negatives of these numbers. They form a set which extends infinitely far in each direction, and consecutive members differ by less than $b - a$. Consequently at least one member of the set must fall between a and b.

Theorem 6: There is a unique function, log, with properties (1), (2), and (3).

Proof: Suppose there are two such functions, log and LOG, and a positive real number a for which $\log(a) < \text{LOG}(a)$. By the preceding theorem there is a rational number r such that

$$\log(a) < r < \mathrm{LOG}(a).$$

But by corollary 5.1, $r = \log(c^r) = \mathrm{LOG}(c^r)$. Therefore we have

$$\log(a) < \log(c^r) = \mathrm{LOG}(c^r) < \mathrm{LOG}(a).$$

Corollary 4.2 now requires that $a < c^r < a$, which is impossible and refutes the supposition that two such functions exist.

In an elementary course the preceding development should perhaps be carried through using base 10 and stopping with corollary 5.1. Next, the graph of $y = \log(x)$ should be sketched. Two points, $(1, 0)$ and $(10, 1)$, are already known, and two more, $(10^{1/4}, 1/4)$ and $(10^{1/2}, 1/2)$, can be approximated. Theorem 4 shows the monotonic nature of the function, and a fair sketch can be made.

Students can now be shown a table of mantissas with the explanation that these are the logarithms of the numbers from 1 to 10. They are next shown how the tables can be extended to the positive numbers less than 1 and to numbers greater than 10 by means of the simple relations:

$$(4) \qquad \log(10N) = 1 + \log(N),$$

and

$$\log(N/10) = \log(N) - 1,$$

which are immediate consequences of properties (1) and (2) and theorem 3.

Characteristics and mantissas can now be defined, and a rule for characteristics determined. Next, the usual computational drill material should be thoroughly covered. For although the student will make fewer algebraic mistakes when taught logarithms in this manner, there is nothing that takes the place of drill to develop computational technique.

Before proceeding to the change-in-base technique and the inverse logarithmic function we need one more fundamental result.

Theorem 7: For every real number y there is a positive real number x such that $y = \log_c(x)$.

Proof: We assume the fact that if a subset of the set of real numbers has an upper bound, then it has a least upper bound. Now let S be the set of all rational numbers such that s belongs to S if $s < y$. Consider now the set c^s. It clearly has an upper bound and therefore a least upper bound, N. We will show that $y = \log_c(N)$. For suppose $\log_c(N) < y$. Then there is a rational number s in S such that $\log_c(N) < s < y$, and therefore $N < c^s$. Thus, N would not be an upper bound for the set c^s, which is contrary to fact, so $\log_c(N)$ cannot be less than y. Next suppose that $\log_c(N) > y$. Then there is a rational number r, not in S, such that $\log_c(N) > r > y$, and therefore $N > c^r$. Since c^r is an upper bound for the set c^s, N is not the least upper bound. Since this too is contrary to fact, we must have $y = \log_c(N)$.

We can now proceed to the relationship of logarithms with different bases and develop a change-in-base procedure. To this end we next prove

Theorem 8: Let k be a positive real number. Then $f(x) = k \log_c(x)$ is a logarithmic function, $\log_{c'}(x)$, with $c' > 1$.

Proof: Since

$$(ab) = k \log_c(ab) = k \log_c(a) + k \log_c(b)$$
$$= f(a) + f(b), \text{ property (1) holds.}$$

If $a > 1$, $\log_c(a) > 0$, and $f(a) = k \log_c(a) > 0$, so property (3) also holds. In view of theorem 7 there is a positive real number, call it c', such that $\log_c(c') = 1/k$. c' will be > 1 since $1/k > 0$.

Now $f(c') = k \log_c(c') = k(1/k) = 1$, and we have property (2) with c' being the base. Since the three properties uniquely determine a logarithmic function, the theorem is established.

Using the same notation as in the theorem just proved, set $x = c$, and we get $f(c) = k \log_c(c) = k$. We can now readily verify

Corollary 8.1. $k = \log_{c'}(c) = 1/\log_c(c')$.

Corollary 8.2.

$$\log_{c'}(x) = \log_{c'}(c) \cdot \log_c(x) = \frac{\log_c(x)}{\log_c(c')}.$$

In view of theorem 7 and the consequences of theorem 4, it is clear that the functional relation, $y = \log_c(x)$, establishes a one-to-one correspondence between the positive real numbers (x) and the set of all real numbers (y). We may therefore consider the inverse function.

The inverse function we shall temporarily call the c-function. Thus, by definition, we shall say that

$$(5) \qquad y = f_c(x),$$

read "y equals the c-function of x," if and only if $x = \log_c(y)$.

From the definition of the c-function it is immediately evident that

$$f_c\,(\log_c(y)) = y, \text{ and } \log_c(f_c(x)) = x.$$

It also follows at once that

$$(6) \qquad f_c(0) = 1, \text{ since } 0 = \log_c(1),$$

$$(7) \qquad f_c(1) = c, \text{ since } 1 = \log_c(c),$$

(8) and for rational r,

$$f_c(r) = c^r, \text{ since } r = \log_c(c^r).$$

Other properties of the c-function are readily obtained, for if

$$y_1 = f_c(x_1), \text{ so that } x_1 = \log_c(y_1), \text{ and}$$

$$y_2 = f_c(x_2), \text{ so that } x_2 = \log_c(y_2),$$

then

$$x_1 + x_2 = \log_c(y_1) + \log_c(y_2) = \log_c(y_1 \cdot y_2),$$

and therefore $y_1 \cdot y_2 = f_c(x_1 + x_2)$, which establishes

Theorem 9: $f_c(a) \cdot f_c(b) = f_c(a+b)$:

Similarly,

$$x_1 - x_2 = \log_c(y_1) - \log_c(y_2) = \log_c(y_1/y_2),$$

so that $y_1/y_2 = f_c(x_1 - x_2)$, and we have

Theorem 10: $f_c(a)/f_c(b) = f_c(a-b)$.

Corollary 10.1. $f_c(-b) = 1/f_c(b)$.

Also, if n is rational,

$$nx = n \log_c(y) = \log_c(y^n), \text{ so that}$$

$$y^n = f_c(nx), \text{ which proves}$$

Theorem 11: $(f_c(x))^n = f_c(nx)$.

In view of the relations (6), (7), and (8) along with theorems 9, 10, and 11, we see that for rational x the c-function has precisely the properties of the exponential function. That is, for rational x,

$$(9) \qquad f_c(x) = c^x.$$

Because of the fact that the inverse logarithmic function is exponential for the rational numbers, we extend it by definition to include all real numbers. Thus by definition, if x is any real number,

$$y = c^x \text{ means } x = \log_c(y).$$

The notation $f_c(x)$ can now be replaced by the simpler exponential notation, and the results rewritten in exponential form. For example, theorem 9 becomes the familiar

$$c^a \cdot c^b = c^{a+b}.$$

Similarly for the other results.

A final advantage of this approach to the logarithmic function is encountered in the calculus. Let $F(x) = \int_1^x dt/t$. It can be shown that $F(x)$ is a logarithmic function by showing that it has the three defining properties (1), (2), and (3). That it has property (1) can be shown as follows:

$$F(ab) = \int_1^{ab} dt/t = \int_1^{a} dt/t + \int_a^{ab} dt/t.$$

Changing variables in the last integral on the right gives

$$t = ay, \; dt = a \cdot dy,$$

and

$$F(ab) = \int_1^{a} dt/dt + \int_1^{b} dy/dy$$

$$= F(a) + F(b), \text{ which is property (1).}$$

It is clear of course that property (3) holds, and there remains to show that there is a base, which we shall call e, such that $F(e) = 1$. To show the existence of such a base, and in fact to approximate its value, consider the trapezoidal areas shown in Figure 1. The one below the curve is tangent to it at the midpoint of the interval so that its area is the same as

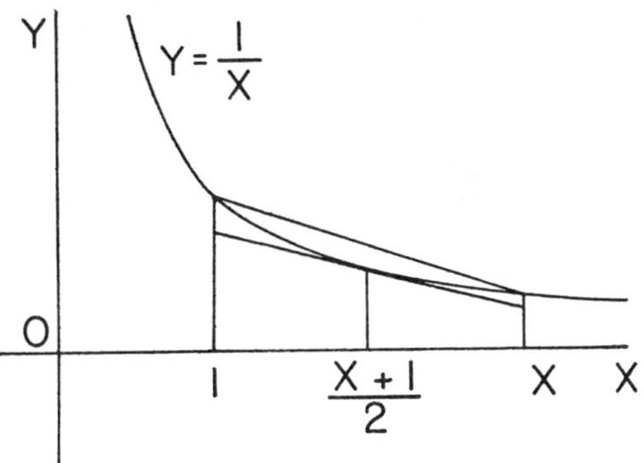

Figure 1

the rectangle with height equal to the mid-point ordinate. Since the curve lies between the two trapezoids, we have the following inequalities:

$$(10) \quad \frac{2}{x+1}(x-1) < \int_1^x dt/t$$

$$< \frac{1}{2}(1+1/x)(x-1) = \frac{x^2-1}{2x}.$$

From these inequalities we find that $F(2) < 1$ and $F(3) > 1$, so that there is an e, $2 < e < 3$, for which $F(e) = 1$. Since this establishes $F(x)$ as a logarithmic function we will now change to the usual notation and write $F(x) = \log_e(x)$, or the more commonly employed $\ln(x)$.

To approximate e let us set $x = 1+h$ and (10) becomes

$$(11) \quad \frac{2h}{2+h} < \ln(1+h) < \frac{h(h+2)}{2(h+1)}.$$

Now by the change-in-base relation given in corollary 8.2,

$$\ln(1+h) = \frac{\log_{10}(1+h)}{\log_{10}e}.$$

Using this in (11) leads to

$$\frac{2(h+1)}{h(h+2)}\log_{10}(1+h) < \log_{10}e$$

$$< \frac{2+h}{2h}\log_{10}(1+h).$$

In this last relation set $h = .001$. In a seven place table of mantissas is found $\log_{10}(1.001) = .0004341$. However, this value contains a rounding error, so we use

$$.00043405 \leqq \log_{10}(1.001) \leqq .00043415.$$

Performing the appropriate arithmetic gives

$$.434266 < \log_{10}e < .434368,$$

and

$$2.7181 < e < 2.7188.$$

We shall clarify an earlier statement concerning postulate (3). To this end we will change postulate (3) and note how the theory is affected. Let the new postulate be

$(3')$ If $a < 1$, then $\log_c(a) > 0$.

It is not difficult to follow through the preceding development and make the necessary changes. We will state the changes required by postulate $(3')$ without proof since they closely parallel the proofs already given.

Theorem 4': If $a < b$, then $\log_c(a) > \log_c(b)$.

Corollary 4'.2. If $\log_c(a) > \log_c(b)$, then $b > a$.

Corollary 4'.3. The base c is less than 1.

Corollary 4'.4. If $a > 1$, then $\log_c(a) < 0$.

The greatest change occurs in theorem 8 as k may be either positive or negative and we may change from a base less than 1 or greater than 1 to another base of either type.

Theorem 8': Let k be any real constant different from zero. Then the function $f(x) = k \log_c(x)$ is a logarithmic function, $\log_{c'}(x)$, and

if $k > 0$ and $c > 1$, then $c' > 1$,

if $k < 0$ and $c > 1$, then $c' < 1$,

if $k > 0$ and $c < 1$, then $c' < 1$,

if $k < 0$ and $c < 1$, then $c' > 1$.

The other theorems and corollaries remain the same. Thus we see that postulate (3) gives a monotonic increasing logarithmic function with a base greater than 1 whereas postulate $(3')$ gives a monotonic decreasing logarithmic function with a base less than 1.

ANTIFREEZE AND EXPONENTIAL GROWTH

By J. F. LEETCH

Bowling Green State University
Bowling Green, Ohio

IN MANY texts the presentation of linear equations includes the problem of replacing a portion (fractional part), x, of the contents of a cooling system by pure antifreeze in order to raise the antifreeze portion from a to b. The result is

$$(1) \qquad x = \frac{b - a}{1 - a},$$

which is valid for $0 \le a < b \le 1$. This method assumes that the operations of drain and refill are performed in sequence.

A similar problem can be phrased in the setting of exponential growth and decay. The frequent illustration of the dilution of a salt solution[1] can be used to solve the antifreeze problem under the assumptions that the draining and refilling are performed simultaneously, with instantaneous mixing, at rate c, with the operation to be completed in time T. Questions that may be raised are (i) What is the nature of a formula for x? and (ii) Which of the two methods is more economical? Solutions to these problems are sketched in the following paragraph. The results, although obtained by routine methods, can be interesting in providing a check on one's intuition.

Let V be the volume of the cooling system, v the volume of water in solution, and

$$s = \frac{v}{V}$$

the portion of water at time t. Then

$$s = e^A e^{\frac{-ct}{V}}$$

for some A. For (i) we are required to find a relationship among x (the portion of the system to be replaced) and the quantities, $c, T, a,$ and b. Letting $s = 1 - a$ when $t = 0$, and $s = 1 - b$ when $t = T$, we find

$$(2) \qquad x = \frac{cT}{V} = \log \frac{1 - a}{1 - b}.$$

From (2) we can see that x does not depend on c or T, that this method is quite uneconomical if

$$\frac{1 - a}{1 - b} \ge e$$

and that it is never more economical than the sequential procedure, since

$$\log \frac{1 - a}{1 - b} \ge \frac{b - a}{1 - a}$$

for those a and b under consideration.

A graphical interpretation of these results is obtained from the observation that for constant x, equations (1) and (2) represent lines through $(1, 1)$ in the ab-

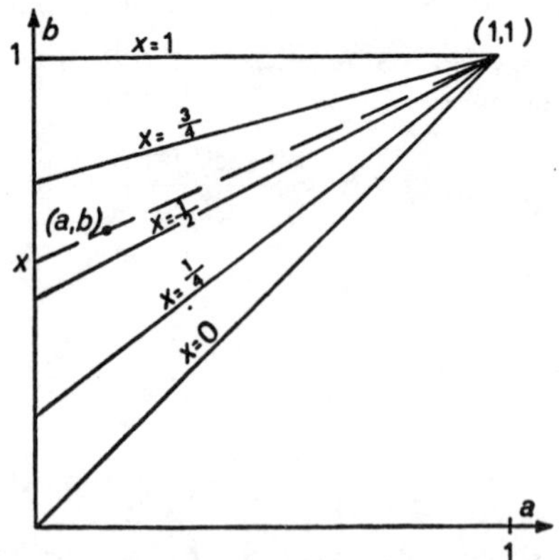

FIGURE 1

THE FAMILY $\{b - 1 = (1 - x)(a - 1)\}$

---

[1] See M. H. Protter and C. B. Morrey, *College Calculus with Analytic Geometry* (Reading, Mass.: Addison Wesley, 1964), p. 479, and A. E. Taylor, *Calculus with Analytic Geometry* (Englewood Cliffs, N.J.: Prentice-Hall, 1959), p. 316.

plane. Several such lines are shown in Figure 1 (sequential method) and Figure 2 (continuous method).

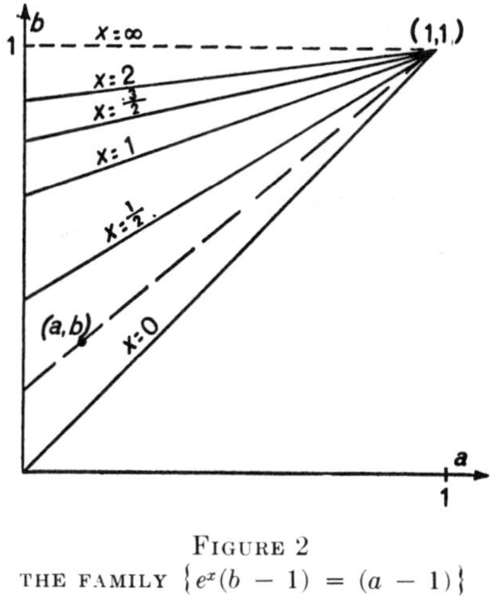

FIGURE 2

THE FAMILY $\{e^x(b - 1) = (a - 1)\}$

From these figures we can observe that the feasible solutions to the original problems are represented by segments in the triangular region $\{(a, b): a \geq 0, a < b \leq 1\}$ and that the "quite uneconomical" solutions to the second problem occur when (a, b) is above the line where $x = 1$ in Figure 2. Furthermore, these figures will suggest a graphic solution to each problem. In the sequential problem (see Fig. 1), plot the point (a, b) corresponding to the specified portions, a and b. The b-intercept of the line through (a, b) and $(1, 1)$ is the required value of x. If sufficient members of the family are shown in Figure 2, one has a network chart with which to plot (a, b) and then estimate the value of x corresponding to the line through (a, b) and $(1, 1)$.

The author is indebted to Professor Fred W. Lott, of the State College of Iowa, for the suggestion of the graphical interpretations.

Bibliography: *Exponential, Hyperbolic, and Logarithmic Functions*

Harper, J. P. Teaching Logarithms. 35 (May 1942): 217–21.
Pedagogical suggestions for teaching logarithms. Includes a method for constructing a common logarithm table.

Schaaf, William L. Logarithms and Exponentials. 45 (May 1952): 361–63.
A bibliographical listing.

Prag, Lewis D. A Device for Teaching Logarithms. 45 (May 1952): 378–80.
Pedagogical suggestions for teaching logarithms.

Schaaf, William L. Note on the Dedication of Logarithms. 50 (April 1957): 295–97.
Quotations from letters by John Napier and his sons concerning the invention of Napierian logarithms.

Yates, Robert C. Logs—Laws of Operation. 51 (December 1958): 608.
A concise expression for the laws of logarithmic operation.

Larsen, Harold D. Pseudo Logarithms. 52 (January 1959): 2–6.
Remarks on some methods used to reduce multiplication to additions and subtractions.

Hartzler, M. E. A Two-and-One-Half-Place Logarithm Table. 53 (March 1960): 183–88.
Remarks on approximating $\log x$ by finding integers m and n such that x^m is approximately equal to 10^n.

Read, Cecil B. John Napier and His Logarithms. 53 (May 1960): 381–84.
Historical remarks concerning the discovery of Napierian logarithms.

Eves, Howard. Naperian Logarithms and Natural Logarithms. 53 (May 1960): 384–85.
Remarks on the relation between Napierian and natural logarithms.

Seebeck, C. L., Jr. The Logarithm Function to the Base e—a Development for High School Seniors. 54 (January 1961): 2–3.
Remarks on a development of natural logarithms using a correspondence between the positive real numbers and the area under the curve $y = 1/x$.

Mucci, Joseph F. Probability and the Radioactive Disintegration Process. 54 (December 1961): 606–8.
Remarks on radioactive disintegration.

Bayer, George L. Setting Up an Approximate Antilog Table. 55 (March 1962): 170–74.
Remarks on constructing a three-place table for antilogs. Antilogs are approximated by inverting key entries in a logarithmic table and then using interpolation.

Jerison, Meyer. Natural Bases of Logarithms. 56 (April 1963): 228–33.
   Remarks concerning logarithms and their bases.

Hacker, Sidney G. Identification of Napier's Inequalities. 63 (January 1970): 67–71.
   Remarks, primarily historical, concerning Napier's inequalities.

Deakin, Michael A. B. A Numerical Approach to Natural Logarithms. 66 (March 1973): 239–42.
   Some pedagogical suggestions concerning natural logarithms.

Limits

Incisive and rigorous thinking are of prime importance for success in mathematics. Too often students, teachers, and even textbook authors are guilty of looseness in definition and usage. Nowhere is this more evident than in dealing with limits. This section emphasizes the care that must be exercised in the verbal and written presentation of the concept of limits.

The article by Huntington provides a good basic introduction. He analyzes several definitions of limit for accuracy and correctness. Written over fifty years ago, this discussion still has pertinence and validity today. Karst uses a dialogue between student and teacher to clarify misconceptions about limits. His script can serve as a guide for a possible student-discovery approach to the teaching of limits.

Limits are a difficult concept for beginning calculus students to understand. The approaches discussed here may aid the instructor in devising a more meaningful and visual presentation than that found in most current textbooks.

RIGHT AND WRONG DEFINITIONS OF A LIMIT.

By Edward V. Huntington.

The following comparison of the correct definition of the *limit of a sequence* with three other definitions, which, although incorrect, are still occasionally to be met with in elementary textbooks, may be of interest to teachers who wish to clarify their ideas not only as to what a limit is, but also as to what it is not.

The six illustrative examples will serve to show that the logical content of each of the four definitions is really different from that of each of the other three.

DEFINITION 1. (Correct.) Suppose we have a sequence of values, u_1, u_2, u_3, . . . , progressing according to any given law; and also a constant quantity c. Then the constant c is called the *limit* of the sequence, *provided* whatever quantity k your opponent may select, *you can always find a stage in the sequence* such that for all values of u beyond this stage, the difference between u and c is less than k.

In other words, whenever your opponent selects a value of k at pleasure, you must be able to find a point in the sequence such that all the values of u beyond this point lie within the range $c - k$ to $c + k$. If, for any selected value of k, the corresponding point in the sequence *cannot* thus be found, then c is *not* the limit of the sequence.

The same definition can be expressed more briefly as follows:

A constant c is called the limit of a variable u, if the difference between the constant and the variable eventually becomes and remains smaller than any pre-assigned quantity k.

DEFINITION 2. (Wrong.) A constant c is called the " limit " of a variable u, if every change in the value of u brings it nearer to c.

DEFINITION 3. (Wrong.) A constant c is the " limit " of a variable u if the difference between c and u can be made less than any pre-assigned quantity, however small.

DEFINITION 4. (Wrong.) A variable u is said to have the " limit " c, if it continually approaches nearer and nearer to c. but never reaches c.

In each of the following illustrative examples, the successive values u_1, u_2 u_3, etc., are represented by points along a line, or rather by the distances to these points from a fixed origin O. The law of variation will in each case be sufficiently clear from the figure, the variable being supposed to run through the values marked 1, 2, 3, etc., in order.

Example 1.

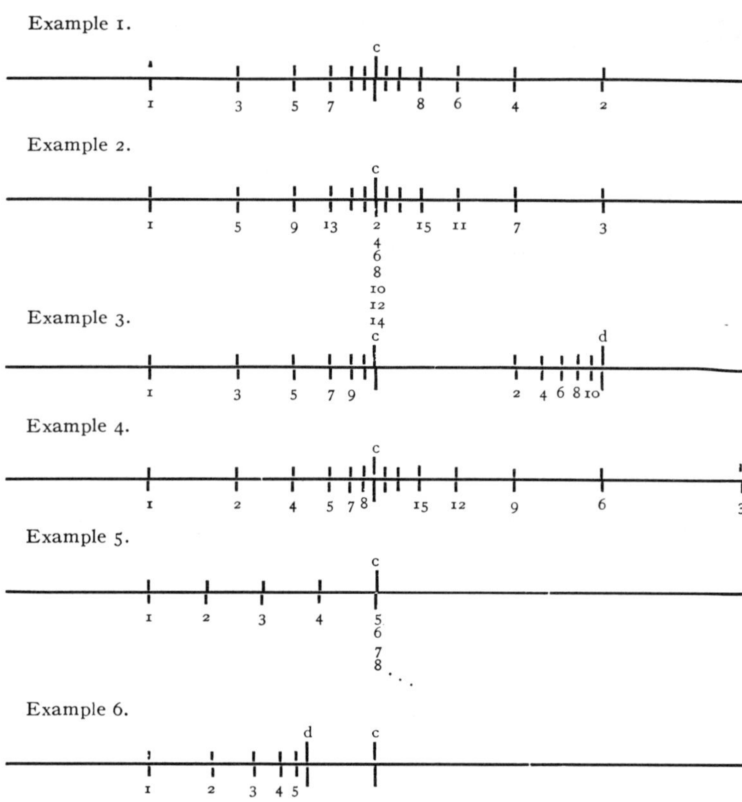

Example 2.

Example 3.

Example 4.

Example 5.

Example 6.

If now we inquire in each of these cases whether the constant c is the "limit" of the variable u, according to each of the four definitions, we obtain the results exhibited in the following table:

Def.	1	2	3	4	5	6
I	Yes	Yes	No	Yes	Yes	No
II	No	No	No	No	Yes	Yes
III	Yes	Yes	Yes	Yes	Yes	No
IV	Yes	No	No	No	No	Yes

It is clear from the definitions that every instance which is "yes" under definition I. will also be "yes" under definition III. With this necessary exception, the table shows that whatever two of the definitions we choose to compare, there is always an instance which is "yes" for one and "no" for the other, and also an instance which is "no" for the first and "yes" for the second. In other words, the four notions of "limit" embodied in the four definitions are absolutely distinct from one another, and should never be confused.

Harvard University,
Cambridge, Mass.

The limit

OTTO J. KARST, *New York University,
New York City, New York.*

*"So you've got to the end of our race-course?" said the Tortoise.
"Even though it does consist of an infinite series of distances?
I thought some wiseacre or other had proved that
the thing couldn't be done?"*—Lewis Carroll.

THERE IS A DEFINITE TREND in secondary school mathematics education toward introducing the elements of the differential and integral calculus in the twelfth year. The teacher who is some years removed from his undergraduate calculus studies is confronted with the problem of refreshing his knowledge of calculus in order to deal competently with the problems of syllabus construction, lesson planning, and actual classroom instruction involving calculus. The objective of this article is to review one of the basic analytic ideas underlying the differential and integral calculus. This is the concept of the limit.

The concept of the limit is central in calculus, since both the derivative and the definite integral are really limits. Every time anyone calculates a derivative or a definite integral, he is finding a limit.

Let us imagine a conversation between a student and his calculus teacher who is trying to clarify some of the student's hazy notions about limits.

STUDENT: The text talks a lot about limits. What is a limit really?

TEACHER: Your use of the word "really" indicates that the *mathematical* concept of limit is somehow inadequately described by the common word "limit" from our English language. This is undoubtedly true, as the technical use of the word implies an understanding of mathematical concepts not at all evident from the word itself. Let me give you a few examples of limits first, and then we'll try to formulate the general concept. Here is a sequence of numbers: 1/1, 1/2, 1/3, 1/4, 1/5, $\cdots$. The general term of the sequence is $1/n$ where n takes on successive positive integral values. What would you say is the limit of the sequence? I don't mean the sum of the terms but rather the limit of the individual term as we go farther out in the sequence.

STUDENT: I'd say the limit was zero, since the successive terms are getting smaller and smaller as n increases.

TEACHER: That's right. The limit is zero, but the mathematician sees much more in the situation than just that. Does any specific term ever *equal* zero?

STUDENT: No, because even if n is, say, one billion, the number 1/1,000,000,000 is not zero, although very close to it.

TEACHER: How close?

STUDENT: Well, if the terms were points on a number scale with units of, say, centimeters, the point for $n = 1,000,000,000$ would be one billionth of a centimeter from zero.

TEACHER: And what about all the terms in the sequence whose n is greater than one billion?

STUDENT: They would be even closer to zero.

TEACHER: Now we have been talking about the limit in a mathematical manner. Suppose I were to name a small number, say, 0.001. What value would n have to be so that the difference between $1/n$ and the limit zero would be less than 0.001?

STUDENT: The value of n would have to be greater than 1000, that is, 1001, 1002, etc. Then 1/1001, 1/1002, and each succeeding term would be closer to zero than 0.001.

TEACHER: In other words, no matter how small a number is chosen, isn't it always possible to find a place in the sequence beyond which all the terms of the sequence are closer to zero than the chosen small number?

STUDENT: Yes. As I begin to see it, the mathematician is not just interested in the limit itself, which in our example is surely zero, but in the closeness to it of something else, in this case $1/n$.

TEACHER: Exactly. The mathematician says that zero is the limit of $1/n$ because $1/n$ can be made as close to zero as anyone desires simply by choosing n large enough. Now note this fine, but very crucial, point: $1/n$ never actually equals zero no matter how large n is.

STUDENT: Do you mean that a variable can have a limit and never actually be equal to the limit?

TEACHER: That is precisely the idea. The actual attainment of the limit value by the variable is not necessary at all.

STUDENT: It seems to me that you go to a lot of trouble to have the variable close to its limit, and then do not care much at all about whether it ever really equals the limit.

TEACHER: Well, that's the whole point. In our example, $1/n$ can't ever equal zero, but it can be as close to zero as we want. As you will see in later developments, particularly when we study the derivative and definite integral, this aspect of the limit concept is very fruitful. What is probably disturbing you is a seeming nonmathematical lack of preciseness in the idea of the limit. Up to now the equality sign has played a dominant role in your mathematical thinking. You are used to saying, for example, $x = 0$, or $x = 4$, which sounds somehow solid and reassuring. Now you

are meeting a more subtle type of mathematical thought. Something does not *equal* something else but merely is very *close* to it. These qualms, which you seem to have, are to be expected. We get comfortably used to an idea, or way of thinking, and then suddenly we are confronted with a situation that requires a new approach. Naturally, you are somewhat resistant to the disturbance of your set thinking pattern. But these are only intellectual growing pains.

STUDENT: Could we look at another specific example of a limit?

TEACHER: Of course. The example I have in mind now arises from a most interesting paradox from the Greeks which indicates that even those formidable thinkers wrestled with the difficulties of the limit concept. Have you ever heard of Zeno's paradox of Achilles and the tortoise?

STUDENT: I've heard of Achilles as a great hero in the Trojan War.

TEACHER: Well, Zeno said that Achilles, great as he was, could not catch the slow tortoise if the tortoise had an initial head start, because if both Achilles and the tortoise started running at the same time, by the time Achilles reached the point where the tortoise started, the tortoise would have crawled to a new position. Then by the time Achilles reached that position, old tortoise would be at still another position. In other words, a faster body can never overtake a slower body because by the time it reaches any position of the slower body the slower body will have moved on.

STUDENT: But that's ridiculous.

TEACHER: Of course it's ridiculous, and the reason for its being so is that we are playing fast and loose with words when we should be treating the situation by mathematical analysis. I think you will admit, though, that old Zeno jolts us a bit by showing how dangerous it is to "verbalize" a problem. Now let's analyze his paradox by making a specific problem of it. Instead of Achilles and the tortoise, we'll consider body F (for fast) and body S (for slow). Suppose that at the time $t=0$, S is at a position one mile ahead of F. Then the distance between them is $d=1$. Now let S move (in a straight line) at the rate of one mile per hour, while F moves at two miles per hour. How long will it take F to get to S's original position, and where will S be then?

STUDENT: It will take F $\frac{1}{2}$ hour, since F's speed is 2 miles per hour and he has to travel 1 mile. By that time S will have gone $\frac{1}{2}$ mile, and the distance between them will therefore be $\frac{1}{2}$ mile.

TEACHER: That's right. Now keep your eye on the distance between them. At the start $t=0$, $d=1$; then at $t=\frac{1}{2}$, $d=\frac{1}{2}$. Now repeat the process. F will take only $\frac{1}{4}$ hour to reach S's second position and by that time S will have moved on another $\frac{1}{4}$ mile. Thus at $t=\frac{3}{4}$, $d=\frac{1}{4}$. A similar calculation will show that at $t=\frac{7}{8}$, $d=\frac{1}{8}$, and so on. In other words, the sequence of successive distances between the moving bodies is

$$1, \quad \frac{1}{2}, \frac{1}{4}, \frac{1}{8}, \cdots \text{ or}$$

$$\frac{1}{2^0}, \frac{1}{2^1}, \frac{1}{2^2}, \frac{1}{2^3}, \cdots \frac{1}{2^{n-1}} \cdots,$$

recognizing that the general term is $1/2^{n-1}$. For example, in the fourth term where $n=4$, we have $1/2^3$.

STUDENT: This is like the first example, because the limit of the nth term is surely zero, since as n gets larger $1/2^{n-1}$ gets smaller. Or, as you would want me to say it, for any chosen number, no matter how small, a certain term in the series can be found such that the difference between it and zero is smaller than the chosen small number.

TEACHER: Right, but you should add that the difference between all *succeeding* terms in the sequence and zero is also smaller than the chosen small number.

STUDENT: But even if the limit of the successive distances between the two bodies is zero, we have not shown that F really overtakes S. In other words, no matter how big n is, there is always a little d left.

TEACHER: But we have shown that the "littleness" of d can be controlled. That is, it can be made smaller than the smallest number you can think of, simply by taking n large enough.

STUDENT: But according to the line of argument, d never does equal zero.

TEACHER: The distance d does not equal zero so long as the time t is less than 1 hour, since at time $t=1-(1/2^{n-1})$, n terms in the d sequence have been generated and $d=1/2^{n-1}$. But time inexorably marches on. When t *equals* 1 hour, as it inevitably will, n becomes infinite, and then we associate the limit value of zero with the distance d. Here, then, is a situation where the mathematical limit has a very real physical significance since in the finite time of 1 hour an infinite number of terms of the d series are generated. Therefore it is the limit of d which we have to consider, and the limit does *equal* zero. Thus Achilles does catch the tortoise in spite of Zeno's objections.

STUDENT: Are there any other examples of limits?

TEACHER: There are many kinds of limits. So far we have discussed only sequences of terms. Let's look now at a limit of a sum of terms, also known as the limit of an infinite series. We can get this out of the Achilles-tortoise type problem which we have had under consideration by asking how far F moves from his position at $t=0$ before he overtakes S.

STUDENT: Well, we should get this by adding up the successive distances that F moves, in other words, the successive d's. Now we want the limit of the *sum* of the d's rather than the limit of the individual terms. That should be $1+\frac{1}{2}+\frac{1}{4}+\frac{1}{8}+\cdots 1/2^{n-1}+\cdots$ and so on indefinitely. But that's easy. I recognize that as a geometric series with a ratio, $r=\frac{1}{2}$, and first term, $a=1$. Therefore, the sum is $s=a/(1-r)$, $=1/(1-\frac{1}{2})$, $=2$. The answer is 2.

TEACHER: If by "answer" you mean the result of inserting numbers into the formula $S = a/(1-r)$ and doing a little arithmetic, then I must admit you have "solved" the problem. But we are not now primarily interested in answers, but rather fundamental concepts. Can you express your ideas in terms of limits?

STUDENT: Let's see. The difference between the sum of the first n terms in the series and 2 can be made smaller than any positive number you can name.

TEACHER: That's fine. You surely have the spirit of the limit concept. Let me try to put it into more precise mathematical language. Let us define $S_n \equiv$ sum of the first n terms of the series $1 + \frac{1}{2} + \frac{1}{4} + \frac{1}{8} + \cdots 1/2^{n-1} + \cdots$. Thus, for example,

$$S_4 = 1 + \tfrac{1}{2} + \tfrac{1}{4} + \tfrac{1}{8}, \text{ and}$$
$$S_5 = 1 + \tfrac{1}{2} + \tfrac{1}{4} + \tfrac{1}{8} + \tfrac{1}{16}, \text{ and}$$
$$S_n = 1 + \tfrac{1}{2} + \tfrac{1}{4} + \cdots 1/2^{n-1}$$

Notice that the series for S_n terminates at $1/2^{n-1}$. Now let me try to translate your previous statement into analytic symbolism. You said in part, "The difference between the sum of the terms of the series and 2," and this I shall write as

$$\left| S_n - 2 \right|.$$

The absolute value signs are used because it is the magnitude of the difference, regardless of its sign, that interests us. You next said that this difference "can be made smaller than any positive number you can name." The mathematician's favorite symbol for this small number is ϵ, the small Greek epsilon. Thus you say

$$\left| S_n - 2 \right| < \epsilon$$

where ϵ is an arbitrarily small, positive number. I'd like to carry this a bit further by saying that for any ϵ you choose, I can find a specific value of n which will make this inequality true. For example, if you choose $\epsilon = 0.1$, then I note that

$$\left| S_4 - 2 \right| = \left| 1\ 7/8 - 2 \right| = .125,$$

and

$$\left| S_5 - 2 \right| = \left| 1\ 15/16 - 2 \right| = .0625.$$

Now $\left| S_6 - 2 \right|$ and every succeeding $\left| S_n - 2 \right|$ will surely be smaller than .0625. Therefore, for all values of $n \geq 5$, the inequality

$$\left| S_n - 2 \right| < 0.1$$

is true.

STUDENT: We are right back to the idea of controlled closeness as in the previous examples.

TEACHER: Right. Now let me try to say this all at once: If

$$S_n = 1 + \tfrac{1}{2} + 1/2^2 + 1/2^3 + \cdots 1/2^{n-1},$$

then for any arbitrarily small positive ϵ, a specific value of n, say N, can be found such that whenever $n \geq N$, then

$$\left| S_n - 2 \right| < \epsilon.$$

STUDENT: The specific value of N depends on the choice of ϵ?

TEACHER: That's right. In fact you can say that N is a function of ϵ or $N = f(\epsilon)$. The actual solution of a limit problem is not just to find the limit, which may be fairly easy, as it was in this case, but rather to find this functional relation between the small number ϵ and some other parameter which controls the range of the variable of problem. In the case we have considered, this other parameter was N, for only when the variable n was greater than or equal to N, was $\left| S_n - 2 \right| < \epsilon$. Therefore the real problem here is to find N as a function of ϵ so that when a specific ϵ is chosen, an N is determined such that when $n \geq N$, then $\left| S_n - 2 \right| < \epsilon$.

STUDENT: That sounds as if it might be a real tough problem.

TEACHER: In general it is, and for the present at least I would like to defer it until we develop the concept of the limit of a function with a continuous independent variable. Let me then put this problem to you: Find

$$\lim_{x \to 2} x^2,$$

where x can have any real value.

STUDENT: I see that x is a continuous variable which makes this different from the discrete cases we have already considered. I am pretty sure that the answer to the problem is 4, but I know by this time that what you want me to discuss is the closeness of x^2 to 4 when x is close to 2. I feel that it would be helpful to tabulate some values.

TEACHER: Good idea. May I suggest the following tabular arrangement?

(1) x	(2) x^2	(3) $\lvert x-2 \rvert$	(4) $\lvert x^2-4 \rvert$

STUDENT: I can see the necessity for the first two columns, but why do we want columns (3) and (4)?

TEACHER: Well, column (3) will measure the closeness of x to 2 and column (4) will measure the closeness of x^2 to 4.

STUDENT: I see. Now I'll substitute some values for x close to 2 and see what happens.

(1) x	(2) x^2	(3) $\lvert x-2 \rvert$	(4) $\lvert x^2-4 \rvert$
2.2	4.84	.2	.84
2.1	4.41	.1	.41
2.01	4.0401	.01	.0401
2.001	4.004001	.001	.004001

TEACHER: That's enough to show some significant trends.

STUDENT: Yes, now I see why you put in columns (3) and (4). When x is within .001 of 2, x^2 is within .004001 of 4. In other words, if you name the number .004001 as the desired closeness of x^2 to 4, I can guarantee that closeness simply by taking x within .001 of 2.

TEACHER: You are almost correct and I hate to mention this, but what if $x=1.999$?

STUDENT: Oh yes! We must consider that, of course. My calculations show that when $x=1.999$, $x^2=3.996001$; $\lvert x-2 \rvert$

$=.001$ and $\lvert x^2-4 \rvert = .003999$, so I am still right! If you make .004001 the closeness criterion for the function, then .001 is the closeness criterion for x regardless of what side of 2 x lies on.

TEACHER: Can you generalize these thoughts? Suppose instead of a specific small number .004001, I merely gave you the general symbol ϵ, implying an arbitrary small positive number. What would you have to find?

STUDENT: I would have to find another small positive number, say, δ, such that whenever x is within δ of 2 the function x^2 would be within ϵ of 4.

TEACHER: In other words, for any arbitrary small positive ϵ, a δ can be found such that when $0 < \lvert x-2 \rvert < \delta$, then $\lvert x^2-4 \rvert < \epsilon$.

STUDENT: Is that what the mathematician has in mind when he writes

$$\lim_{x \to 2} x^2 = 4?$$

TEACHER: Exactly.

STUDENT: One thing still puzzles me. If the problem "Find

$$\lim_{x \to 2} x^2 ''$$

were put to you, you wouldn't actually go through the process of constructing the tables or investigating the ϵ, δ relation; you would just substitute 2 for x in x^2 and get 4, wouldn't you?

TEACHER: Yes.

STUDENT: Isn't that being a little intellectually dishonest to find the limit by letting x equal 2 when the whole philosophy of the concept of the limit is that x can't equal 2?

TEACHER: That is a sharp question which shows that you are beginning to develop a healthy critical state of mind. I'll try to answer you by showing that there are some situations where the limit can be found by direct substitution quite legitimately, and other situations where it is impossible. As a counterexample consider this problem: Find

$$\lim_{x \to 0} \frac{\sin x}{x} \cdot$$

STUDENT: When $x = 0$ is substituted in $\sin x/x$ we get 0/0, which is meaningless. Yet I recall that this limit equals 1.

TEACHER: Yes, and it is found by a geometric proof. How about this one,

$$\lim_{x \to 2} \frac{x^2 - 4}{x - 2}?$$

STUDENT: Again indeterminate.

TEACHER: Now try

$$\lim_{x \to 2} \frac{4}{x - 2}$$

by direct substitution.

STUDENT: This would be 4/0, which is meaningless.

TEACHER: Can you see how each of these counter examples differs from

$$\lim_{x \to 2} x^2,$$

which can be evaluated by direct substitution?

STUDENT: Well, something strange happens to the function at the limit; it either is indeterminate or infinite.

TEACHER: The mathematician would say that the function does not then *exist*. Under those circumstances the limit can't be found by direct substitution.

STUDENT: Then so long as the function exists at $x = a$,

$$\lim_{x \to a} f(x)$$

can be found by direct substitution of $x = a$ in $f(x)$? In other words,

$$\lim_{x \to a} f(x) = f(a)?$$

TEACHER: The existence of the function is only a *necessary* condition. There are still other conditions which have to be met before you can safely find the limit by substitution. *All* of these conditions together establish the continuity of the function at $x = a$. If the function is continuous at $x = a$, then

$$\lim_{x \to a} f(x)$$

does equal $f(a)$, and one can use substitution of $x = a$ to find the limit.

STUDENT: What are these conditions for continuity?

TEACHER: You can determine if a function $f(x)$ is continuous at $x = a$ by determining (1) if $f(x)$ is defined at $x = a$, that is, if $f(a)$ exists; and (2) if

$$\lim_{x \to a} f(x)$$

exists. If $f(a)$ exists and

$$\lim_{x \to a} f(x)$$

exists, and if

$$\lim_{x \to a} f(x) = f(a),$$

then you have shown that $f(x)$ is continuous at $x = a$. If it happens that $f(x)$ is not defined at $x = a$, or that

$$\lim_{x \to a} f(x)$$

does not exist, or that, in case both exist, they are different numbers, then you have determined that $f(x)$ is discontinuous at $x = a$. I realize that this is pretty technical, and perhaps some day we should discuss the concept of continuity in greater detail. For the present we can say very roughly that a function is continuous at a point with abscissa $x = a$, if its graph has no breaks there. However, the really rigorous definition of continuity proceeds as I first indicated.

STUDENT: There is one more aspect of limits that I would appreciate your showing me. Would you work out the functional relationship between ϵ and δ for a specific problem?

TEACHER: Surely. The very heart of any limit problem is to know what δ one needs in $|x - a| < \delta$ to meet the requirements of any ϵ in $|f(x) - L| < \epsilon$. I'm using L as the symbol for the limit of the function $f(x)$, as $x \to a$. For a specific problem, consider the old stand-by:

$$\lim_{x \to 2} x^2 = 4.$$

Our problem now is not to find the limit 4. It is easy to see that x^2 is close to 4 when x is close to 2. What we want to examine is the relation between $|x^2 - 4| < \epsilon$ and $|x - 2| < \delta$. In other words, to find δ as a function of ϵ, so that when ϵ is specified as the desired closeness of x^2 to 4, δ will be determined as the necessary closeness of x to 2.

STUDENT: In other words, if someone gives you some value for ϵ like .001, then the function of ϵ could be used to find the numerical value of δ, which would be the range around 2 within which x would have to lie to make x^2 lie within .001 of 4.

TEACHER: That puts it very well. Now see if you can follow the analysis where I use the general symbol ϵ instead of .001.

$$|x^2 - 4| < \epsilon$$

can be written

$$-\epsilon < x^2 - 4 < \epsilon,$$

since $x^2 - 4$ can be either positive or negative but not larger in magnitude than ϵ. This inequality is satisfied by the same values of x if 4 is added throughout. Therefore, we have

$$4 - \epsilon < x^2 < 4 + \epsilon.$$

Taking the positive square root throughout, we get

$$\sqrt{4 - \epsilon} < x < \sqrt{4 + \epsilon}.$$

This inequality expresses the range of positive x consistent with

$$|x^2 - 4| < \epsilon.^*$$

Now, surely,

* Actually, this range only holds if $0 < \epsilon \leq 4$. When $\epsilon > 4$, we have $-\sqrt{4 + \epsilon} < x < \sqrt{4 + \epsilon}$. This, however, leads to the same ϵ, δ relation as the inequality considered.

$$\sqrt{4 - \epsilon} < 2 < \sqrt{4 + \epsilon},$$

since ϵ is positive. All we have to do now is to see which of $\sqrt{4 - \epsilon}$ and $\sqrt{4 + \epsilon}$ is closer to 2. Then this range will equal δ. A little thought will tell you that

$$\left| 2 - \sqrt{4 - \epsilon} \right| > \left| 2 - \sqrt{4 + \epsilon} \right|,$$

since $\sqrt{4 - \epsilon} < \sqrt{4 + \epsilon}$. Therefore, $\sqrt{4 + \epsilon}$ is closer to 2, and we finally get $\delta = \sqrt{4 + \epsilon} - 2$ as the desired functional relation.

STUDENT: That was rough.

TEACHER: Yes, unfortunately each limit problem presents its own unique difficulties in finding the ϵ, δ relation.

STUDENT: I may not be able to reproduce that development myself, but I'm pretty sure I know its meaning.

TEACHER: What is that?

STUDENT: Well, if ϵ is given as, say, .01, we can now calculate δ exactly as $\delta = \sqrt{4.01} - 2 = .00250$, and we can then say that if $|x - 2| < .00250$, then

$$|x^2 - 4| < .01.$$

TEACHER: That's it. We now have the complete picture about

$$\lim_{x \to 2} x^2 = 4.$$

We not only know that the limit is 4, but with the relation $\delta = \sqrt{4 + \epsilon} - 2$ we also know exactly what δ to use for a given ϵ to be able to say that when $|x - 2| < \delta$, then $|x^2 - 4| < \epsilon$.

STUDENT: What is there to learn next about limits?

TEACHER: Well, the derivative and the definite integral are both limits. If you'll drop in again sometime, we can see how the limit concept is applied to those two basic ideas of calculus.

Bibliography: *Limits*

Roe, E. D., Jr. A Generalized Definition of Limit. 3 (September 1910): 43–48.
  Remarks are of historical interest only. The definition of limit that is given is the fore-runner of the $\in$, δ concept.

Stromquist, Carl Eben. A Geometric Illustration of Limits. 11 (September 1918): 34–35.
  Counterexamples to the statement that a variable cannot reach its limit value.

Sanford, Vera. Teaching Incommensurables. 14 (March 1921): 147–50.
  Some geometric illustrations of the limit concept.

Jackson, Dunham. The Notion of Limit. 17 (February 1924): 72–77.
  The concept of limit is explained by using the sum of an infinite series and also by finding the circumference of a circle.

Perel, W. M. The Teaching of Mathematical Infinity. 51 (April 1958): 263–66.
  Remarks concerning some pitfalls in an indiscriminant use of the word *infinity*.

Esposito, John N. Infinite, Finite, Infinity. 53 (May 1960): 395–97.
  Some pedagogical suggestions for presenting finite and infinite sets as well as the concept of infinity.

Hight, Donald W. The Limit Concept in the SMSG Revised Sample Textbooks. 57 (April 1964): 194–99.
  A discussion of the limit concept from the geometric viewpoint (i.e., approximating the circumference of a circle), the $\in$, δ approach, and using the tangent to the graph of a function.

Rotando, Louis M. Continued Square Roots. 58 (October 1965): 507–8.
  Remarks concerning the limit of a sequence. Numerical examples of sequences of real numbers are given as well as sequences of imaginary numbers that have real limits.

5

Differentiation

Calculus texts all contain detailed chapters on differentiation formulas, geometric interpretations of derivatives, maximization and minimization problems, and general applications. It would seem that nothing new might be added. The articles selected here, however, present innovative approaches to these questions that can spark further interest and insights in the classroom.

The first part of this section deals with the teaching of differentiation. The concept of tangent line to a curve is treated by Andree. Several pseudodefinitions are proposed and exploded. Holden describes a graphing technique to stimulate interest in learning the chain rule.

Applications of differentiation and differentials are the central theme of the second part. Ogilvy analyzes the accuracy of square root extraction by means of differentials. Differentiation techniques are applied by Stretton to describe the behavior of a particle moving in a straight-line tunnel through the earth.

The third part focuses on problems dealing with maxima and minima. Adler considers several such problems and in each instance presents both a calculus as well as a geometric treatment. Next, Dorn shows how computers can be used effectively in promoting the study of maxima and minima. The use of directional derivatives in finding the extrema of a function of two variables is discussed by Stretton.

Much of the material in this section is directly adaptable to classroom use. It can also stimulate both teacher and student to discover implications and applications of differentiation concepts.

What is a geometric tangent?

RICHARD V. ANDREE, *University of Oklahoma, Norman, Oklahoma.*
Is your *definition adequate?*

Do YOU KNOW the meaning of the phrase "line tangent to a curve"? Before you read on, take a minute to write down your definition of a "line tangent to a curve." Your definition will be an important part of the article, so please write it out now.

Before considering a definition which has been found quite satisfactory, let us examine some of the pseudo-definitions which students often suggest.

Pseudo-definition 1: "A tangent is a line perpendicular to the radius (at its extremity)."

Even in the case of a circle, the definition is inadequate since the line segment "radius" has two extremities. This difficulty can be overcome, but the definition is still inadequate since it is valid only for circles. (Readers who dimly recall a course in calculus may grasp at a straw called "radius of curvature." This is inadequate, since the definition of radius of curvature involves the idea of tangent line, and circular reasoning results.)

Pseudo-definition 2: "A tangent is a line which touches the curve at only one point."

The sketches on this page demonstrate that this description is inadequate:

Figure 1
Line *L* touches curve *C* at only one point *P*, but *L* is *not* a tangent line.

Figure 2
Line *L* is a tangent to curve *C* at point *P*, but it also crosses curve *C* at at least three other points.

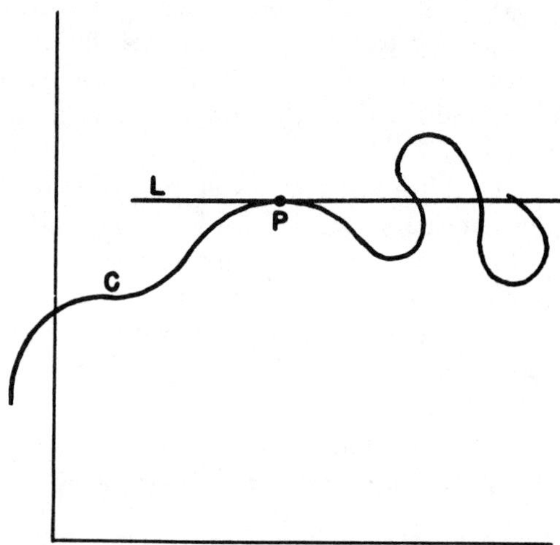

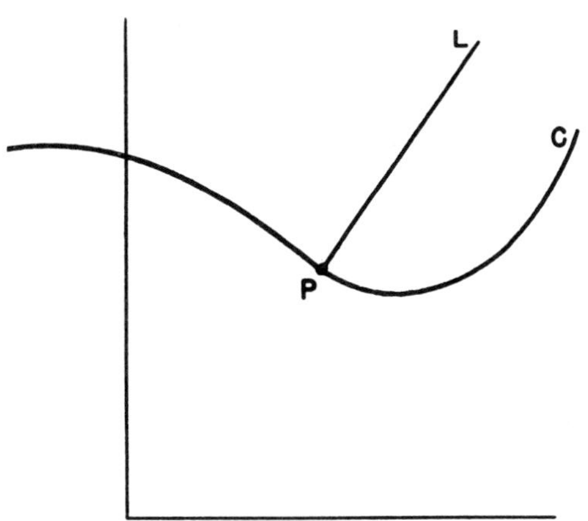

Figure 3

Line segment L touches curve C at point P, but does not cross it; however, L is *not* tangent to C at P.

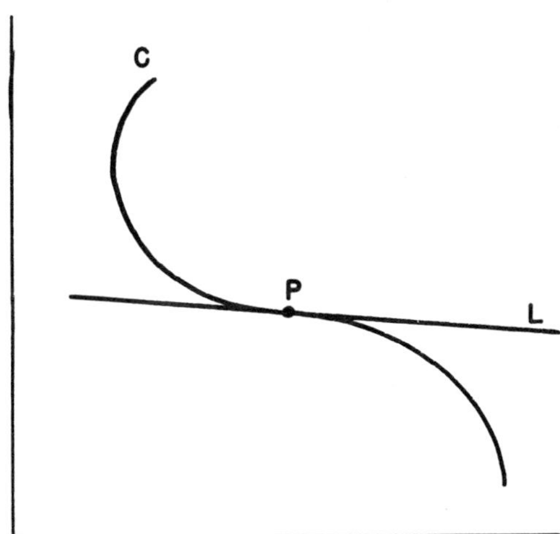

Figure 4

Line L is tangent to curve C at point P, but L also crosses C at P.

Pseudo-definition 3: "A tangent is a line which touches the curve, but does not cross it."

Figures 3 and 4 indicate that the proposed statement does not provide an adequate definition. Furthermore, the statement involves the ungeometric concept of "touching but not crossing," which can hardly be condoned in either synthetic or analytic geometry. Some readers will (rightly) object that the line segment terminating at P in Figure 3 is not a line, but a half line. However, Figure 4 shows a line which is tangent to the curve, and also crosses the curve *at the point of tangency.*

Pseudo-definition 4: "A line L is tangent to a curve C at a point P if, and only if, P lies on both L and C and it is possible to construct a small circle with center P such that C and L have no point in common inside the circle except the point P."

This seems a bit complicated, but if it were valid, it would not be too complicated to be useful. However, it too is neither necessary nor sufficient. It is clearly not sufficient, since any line crossing the curve (Figure 1) satisfies pseudo-definition 4. Furthermore, the condition is not even necessary, although a counterexample is somewhat more difficult to arrange. The

curve consisting of that portion of $y = \sin\ (\pi/x)$ which lies between $x = -1$ and $x = 1$, combined with the half circle $y = +\sqrt{1-x^2}$, provides one counterexample, since the line $y = 1$ is tangent to the curve at $(0, 1)$, but in any small circular region about $(0, 1)$ the line $y = 1$ and the **closed curve** have infinitely many points in common. Other counterexamples may also be found. However, this is unnecessary since Figure 1 proves the definition inadequate.

In ordinary classroom teaching, the author usually has the students suggest possible definitions, and then gives counterexamples to show why the suggested definitions of "line L tangent to a curve C" are inadequate (which they usually are). After perhaps ten minutes the class is about ready to concede that perhaps they are not able to define "line tangent to a curve," although most of them still feel (perhaps justifiably) that they know what one is. This is a crucial point at which it is possible to introduce a valid definition. It is my custom to explain to the students that, before continuing with the study of mathematics, it is necessary to have a valid definition of tangent line, since much of the work in mathematics (I avoid the word calculus) depends upon this concept,

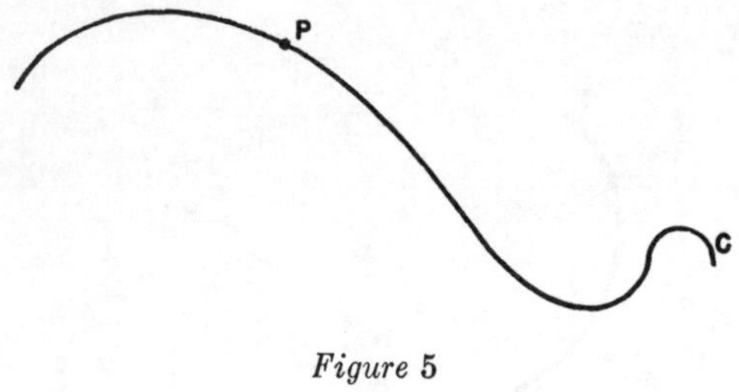

Figure 5

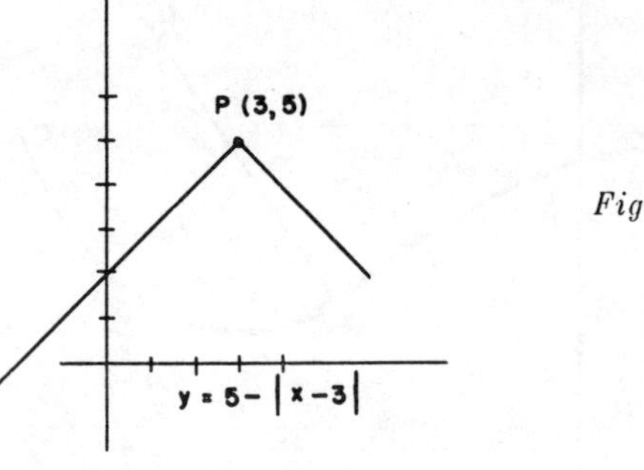

Figure 7

and that experience suggests that the following definition meets the requirements.

First, consider a curve C on which the fixed point P (the proposed point of tangency) has been located. Next let S_1, S_2, S_3, $\cdots$ be any points on C, except P, and consider the secant lines PS_1, PS_2, PS_3, PS_4, $\cdots$.

If the sequence of points S_1, S_2, S_3, S_4, $\cdots$, each distinct from P, is so chosen (on C) that the sequence of *arc lengths*, PS_1, PS_2, PS_3, PS_4, $\cdots$ approaches length zero [i.e., $\lim_{n\to\infty}$ (length of arc PS_n) $= 0$], then the sequence of secants PS_n will have the tangent line at P as limiting position, if such a tangent line exists.

Definition: The tangent line PT to a curve at a fixed point P on the curve is the limiting position of the sequence of secant lines PS_n, where the points S_n are so chosen on the curve that the sequence of (lengths of arcs PS_n) approaches zero, *providing this limiting position exists and is unique.*

After some boardwork with a ruler, it is time to consider the possibility of a (continuous) *curve which fails to have a tangent* at a given point on the curve. Figure 7 serves admirably.

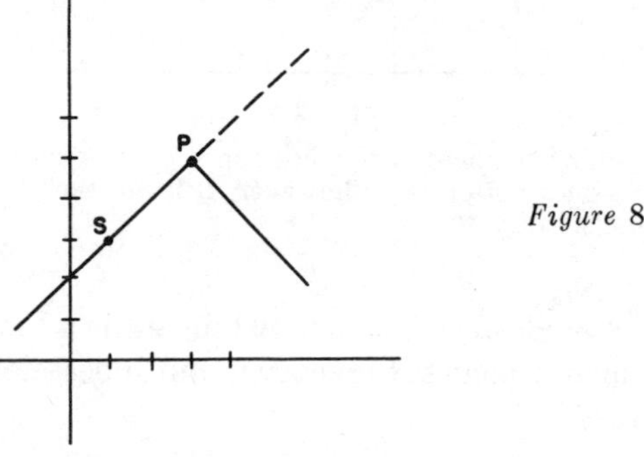

Figure 8

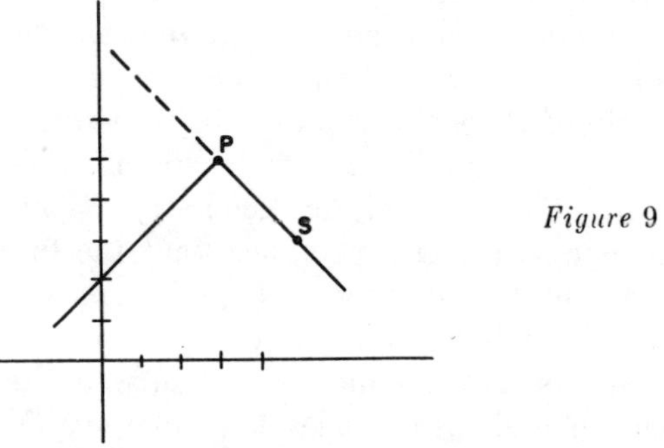

Figure 9

Figures 8 and 9 show two possible positions of S, and the dotted lines show the (one-sided) limiting positions of secants through P which contain points S; on the

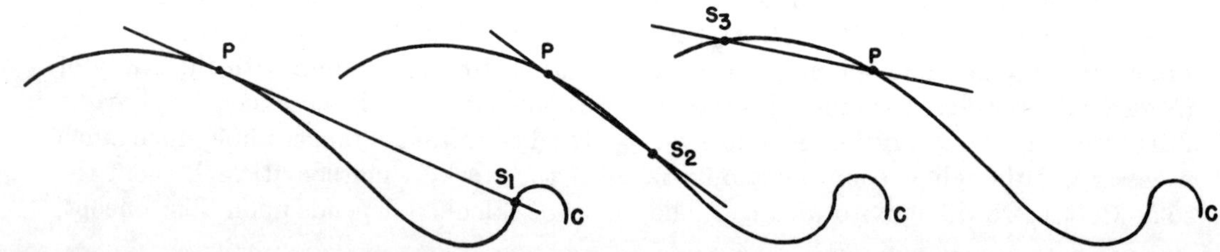

Figure 6

Three possible positions of secant PS.

curve on each side of P. Since the two (one-sided) limiting positions are not the same (i.e., not unique), *the given curve has no tangent line at point P.* It should be pointed out to the students that the definition requires uniqueness, and that no tangent line exists in this case.

Teachers who are including analytic geometry in their tenth-grade course may wish the equation of a curve having no tangent line at a given point. The broken line sketched in Figure 7 has equation $y = 5 - |x-3|$ where the symbol $|x-3|$ means the absolute value (value with positive sign) of $(x-3)$. Point P has co-ordinates (3, 5).

If you have not considered the many advantages of including co-ordinate (analytic) geometry in your tenth-grade class, you should do so. Many teachers are enthusiastic. The current New York State Syllabus specifies portions of analytic geometry to be included in the tenth-grade course. It is your duty as a teacher to learn what is going on, and then form your own opinion. Learn what others are doing before forming a firm opinion. Don't say, "we can't do that here." You don't know until you try.

The presentation of the meaning of tangent line suggested in this note has been tried in numerous classes. Students grasp it readily once it has been explained. Furthermore, it provides an opportunity to discuss what an adequate definition is, and why it is needed. This definition provides the first directed thinking on the matter of limits which the student encounters. Don't handicap your students by failing to discuss geometric limit.

The author invites correspondence from interested teachers.

EDITORIAL COMMENT: If you are willing to grant that a line is tangent to itself (which it is), then a simpler counterexample to pseudo-definition 4 is obtained by considering C as a line, P any point on that line; the tangent line L to C at P is C itself. Inside any circle with P as center, C and L have infinitely many points in common.

PROVIDING MOTIVATION FOR THE CHAIN RULE

By LYMAN S. HOLDEN

Southern Illinois University
Edwardsville, Illinois

THE purpose of this article is to describe how a certain graphing technique can be used to increase the heuristic appeal of a general formula from the differential calculus. The formula to which reference is made is the formula for the derivative of the composition of two functions (the chain rule). We shall interpret this rule in relation to the graph of the function that is the composition of two functions. Using the geometric interpretation of the derivative as the slope of a curve, it will be possible to provide improved motivation for the rule.

The beginning calculus student sometimes has difficulty understanding both the chain rule and its proof. By establishing an association between the rule and the graphs of the functions involved it is possible to give geometric meaning to the rule. The expressions involved in the deduction of the rule as well as the derivatives that appear in its final formulation can be visualized. In this way the student can be shown that the rule is a reasonable one and that the rule makes sense intuitively. The writer believes that after an appropriate buildup the student can even be led to "discover" the statement of the chain rule for himself. It is suggested that the content of this article represents a

natural sort of extension or application of a graphing technique to the problem of providing motivation for the rule.

To interpret the chain rule we will employ a technique for graphing composite functions that has been published only recently. Techniques for graphing composite functions have been known for some time, but to the writer's knowledge all of these construct the graph as a curve in three-space.[1] In an article that appeared in THE MATHEMATICS TEACHER Haddock and Hight describe a geometric construction of the graph of a composite function as a curve in the plane.[2] This construction is particularly well suited to our purpose here.

We first discuss the geometric technique for constructing the graphs we need and follow this with the interpretation of the differentiation formula in connection with these graphs.

Let us recall how composite functions are defined. Suppose $y = f(u)$ and $u = g(x)$ where each of y, u, and x is a real variable. We can define a new function

---

[1] Methods for representing composite functions as curves in three dimensions can be found in the following references: William H. Echols, *An Elementary Text-book on the Differential and Integral Calculus* (New York: Henry Holt & Co., 1902), p. 453. Kurt Kreith, "A Geometric Construction of Composite Functions," *American Mathematical Monthly*, LXIX (April 1962), 293–94.

[2] Glen Haddock and Donald W. Hight, "Geometric Techniques for Graphing," LIX (January 1966), 2–5.

$y = F(x)$, called the composition of f with g, in this way:

$$y = F(x) = f[g(x)]$$

where

$$D_F = \{x \mid x \in D_g \land g(x) \in D_f\}.$$

Some authors define the composition of two functions without introducing an intermediate variable u. The notation suggested here, which employs a variable u that is at the same time a dependent and an independent variable, has been adopted advisedly. From experience in the classroom it appears that this notation is the one best suited for later reconciliation with the statement of the chain rule as found in a typical calculus textbook.

We now describe the method for constructing the graph in the plane of the function that is the composition of two given functions. It is this construction that will be used to give pictorial meaning to the chain rule. Notice that only a straightedge is required to obtain the desired graph. The procedure outlined below is essentially that of Haddock and Hight, but the notation has been slightly modified.

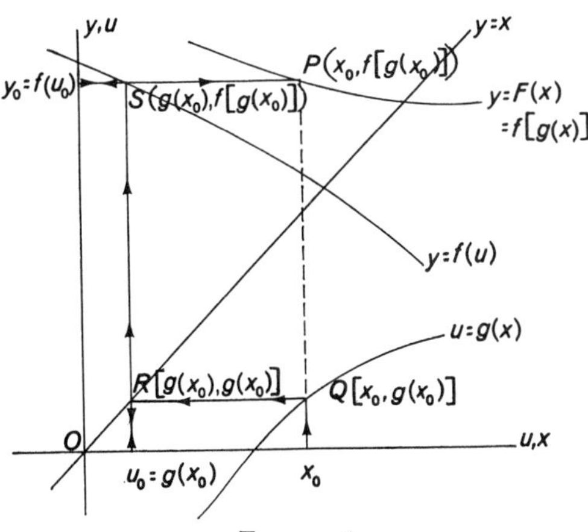

FIGURE 1

Suppose we are given two functions, $y = f(u)$ and $u = g(x)$, and that we seek a pictorial representation of $y = F(x) = f[g(x)]$ (see Fig. 1). It is convenient to think of both the horizontal and the vertical axes of reals as playing a double role. Thus we graph $y = f(u)$, thinking of the vertical axis as the y-axis and the

horizontal axis as the u-axis, and then graph $u = g(x)$, with the vertical axis as the u-axis and with the x-axis playing its customary role. The line $y = x$ is important in this technique, and it is shown in Figure 1. Now, given a point $x = x_0$ on the x-axis, $x_0 \in D_F$, we wish to locate the point $P(x_0, f[g(x_0)])$. This is accomplished as follows. From the point $Q(x_0, g(x_0))$ we draw a horizontal line to intersect the graph of $y = x$ at the point $R(g(x_0), g(x_0))$. A vertical line through R will intersect $y = f(u)$ in the point $S(g(x_0), f[g(x_0)])$, and a horizontal line through S will intersect the line $x = x_0$ in the desired point $P(x_0, f[g(x_0)])$, which is on the graph of $F(x) = f[g(x)]$. The arrows in Figure 1 indicate the path to be followed. In a sense, the graph of $g(x)$ and the line $y = x$ serve to map real numbers on the horizontal axis back into that axis. The procedure carries x_0 to the curve $g(x)$, to the line $y = x$, and then back to the horizontal axis at the point $u_0 = g(x_0)$. Then u_0 is carried to the point $y_0 = f(u_0)$ on the vertical axis, and the last horizontal line marks off this ordinate on the line $x = x_0$ at the point P.

The writer tried this technique not long ago with a class of first-quarter calculus students, and no great difficulty was encountered. Using coordinate paper, these students were able to graph with very good accuracy the function $y = \cos x^2$ from the graphs of $y = \cos u$ and $u = x^2$. The construction of such graphs is recommended by the fact that the student gains insight into the nature of composite functions, and by the fact that he sees $y = F(x) = f[g(x)]$ as a new function that is distinct from, but related to, the two given functions.

Now the theorem from the differential calculus that we intend to relate to this graphing technique is this:

Theorem (the chain rule). Let $y = f(u)$ be a differentiable function of u, and $u = g(x)$ be a differentiable function of x. Then the composite function $y = F(x) = f[g(x)]$ is a differentiable function of x, and the derivative of y with respect to x is the product of the derivative of y with respect to u and the derivative of u with respect to x; i.e.,

$$F'(x) = f'(u) \cdot g'(x)$$

or

$$\frac{dy}{dx} = \frac{dy}{du} \cdot \frac{du}{dx}.$$

Proof. The proof of this theorem is complicated by the fact that for certain functions $u = g(x)$ and certain points x in D_F when we give x an increment $\Delta x \neq 0$ the corresponding increment in u, Δu, may be zero. We therefore make the assumption that for each x in D_F there corresponds a number $\sigma > 0$ such that for all Δx satisfying $|\Delta x| < \sigma$ the corresponding values of Δu are never equal to zero.[3] Under this assumption we can write the identity

$$\frac{\Delta y}{\Delta x} = \frac{\Delta y}{\Delta u} \cdot \frac{\Delta u}{\Delta x}, \quad |\Delta x| < \sigma.$$

Taking limits, and noting that when $\Delta x \to 0$ then $\Delta u \to 0$, we have:

$$\lim_{\Delta x \to 0} \frac{\Delta y}{\Delta x} = \lim_{\Delta x \to 0} \left(\frac{\Delta y}{\Delta u} \cdot \frac{\Delta u}{\Delta x} \right)$$

$$= \left(\lim_{\Delta u \to 0} \frac{\Delta y}{\Delta u} \right) \cdot \left(\lim_{\Delta x \to 0} \frac{\Delta u}{\Delta x} \right),$$

or

$$\frac{dy}{dx} = \frac{dy}{du} \cdot \frac{du}{dx}.$$

Three Categories

David Gans has written a perceptive article in the *American Mathematical Monthly* that bears directly on the matter of assuming that $\Delta u \neq 0$ when $\Delta x \neq 0$.[4] He classifies textbook proofs of the chain rule into three categories. A proof in Category A consists of writing the identity

$$\frac{\Delta y}{\Delta x} = \frac{\Delta y}{\Delta u} \cdot \frac{\Delta u}{\Delta x},$$

taking limits to obtain the result, and no mention is made of the complicating fact that Δu may be zero when Δx is not zero. A proof in Category B is like the above but includes a brief remark that Δu must not be zero. Gans says that

this proof [Category B] can be regarded as an improvement over the preceding [Category A] only if it is granted that it is better to leave a student mystified by a difficulty than ignorant of it.[5]

A proof in Category C is rigorous and complete and uses an ϵ-procedure of some sort. Of proofs in Category C Gans says this:

However, it is hard to see how these comparatively sophisticated proofs can be understood by any but a very small minority of students when it is realized that the latter are usually in contact with the calculus for only about a month when the chain rule is presented to them.

The last paragraph of this article contains Gans's recommendations for giving a proof that is both rigorous and understandable. He says:

If x_1 is the value of x under consideration, and u_1 the corresponding value of u, all that is now necessary [for a rigorous, understandable proof] beyond what is presented in textbooks is to agree to consider only functions $u(x)$ for which Δu is not zero if x is sufficiently close to x_1 but unequal to it. That is, we state and prove the chain rule only for such functions. Thus we only exclude functions taking on the value u_1 infinitely many times in the neighborhood of x_1. Except for the trivial case when $u(x)$ is constant in this neighborhood, such excluded functions never occur in first year work, and but rarely in the following years. The student should be told that the chain rule also applies to these functions but that the proof of this fact is best delayed until such time as he works with them.[6]

Thus Gans advocates a proof in Category D, say, that retains the simplicity of a proof in Category A by excluding certain unusual functions and avoids the possibly excessive rigor of a proof in Category C. A proof in Category D would include an expanded explanation

---

[3] If it is the case that there are arbitrarily small values of Δx distinct from zero for which $\Delta u = 0$, then the theorem still holds but the proof must be modified.

[4] David Gans, "On Proving the Chain Rule," *American Mathematical Monthly,* LXII (1955), 115–16.

[5] *Ibid.*, p. 115.

[6] *Ibid.*, p. 116.

of the difficulty when Δu is zero. It would include examples of elementary functions $u(x)$ (sin x, for example) where Δu can be zero when Δx is not zero, but where Δu cannot be zero if Δx is sufficiently small.

It is felt that the chain rule as stated above is consonant with Gans's recommendations.

To Encourage Understanding

By making a slight sacrifice in the generality of the theorem we obtain a brief and simple proof. But brevity of exposition does not assure understanding on the part of the student. Experience with teaching this theorem in the classroom shows that it is a difficult result for the student to apply. Often the symbols involved are largely devoid of meaning, and the student learns to perform differentiations in a mechanical sort of way. As teachers and textbook writers, we make almost no effort to have the chain rule appear reasonable to the learner. This may be because we don't know how to make it appear reasonable or because it is deemed unnecessary to do so.

The writer suggests that the chain rule can be made plausible by employing the geometric technique for graphing composite functions that was outlined above. We consider the following sequence of ideas, and refer to Figure 2 for pictorial assistance. The language used refers to graphs as shown in Figure 2. For a different orientation the words up, down, left, and right would be modified to conform to the arrangement of the graphs. It is hoped that by following this sequence of ideas the student will benefit in two ways: the symbols he is working with will have increased meaning, and the result itself will appear plausible. The *basic clue* that serves to give direction to the investigation is this: that the derivative of the new function is expressible in terms of the derivatives of the functions of which it is made up.

Figure 2 shows first of all the graphs of three functions, $u = g(x)$, $y = f(u)$, and

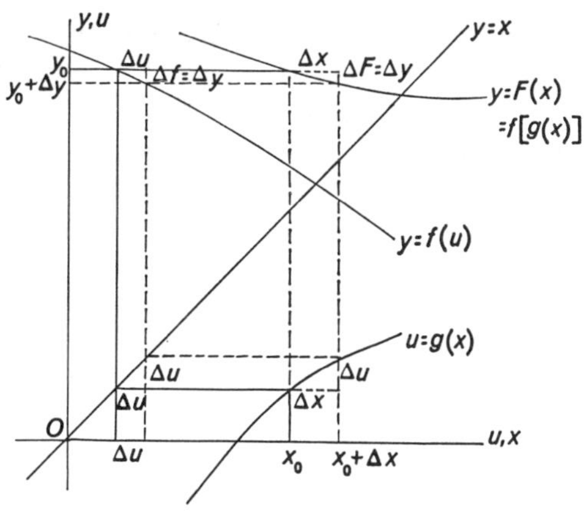

FIGURE 2

$y = F(x)$. We seek an expression for $F'(x)$ in terms of the derivatives $f'(u)$ and $g'(x)$ of its component functions. We work at a point x_0 in D_F and thereby emphasize the fact that the chain rule is a statement about a pointwise property of $F(x)$, a fact that is easily lost sight of by the student.

A change in x_0 by the amount $\Delta x \neq 0$ induces a chain of related changes. It generates a change in u, shown in Figure 2. If

$$u_0 = g(x_0)$$

and

$$u_0 + \Delta u = g(x_0 + \Delta x),$$

then

$$\Delta u = g(x_0 + \Delta x) - g(x_0).$$

In view of the way the geometric technique for graphing works, x_0 maps up to $g(x)$, across to the line $y = x$, where it is reflected toward the point $u_0 = g(x_0)$ on the x-axis. Similarly, $x_0 + \Delta x$ maps up, across, and down to the point $u_0 + \Delta u = g(x_0 + \Delta x)$. One can imagine the line segment above $x_0 + \Delta x$ which represents Δu as being carried left to the line $y = x$, rotated 90° clockwise, and subsequently appearing as an interval on the horizontal axis. Now, referring to the graph of $y = f(u)$, we can visualize the change in y that is generated by this Δu. u_0 maps up and across into $y_0 = f(u_0)$ and $u_0 + \Delta u$ maps into $y_0 + \Delta y = f(u_0 + \Delta u)$. Thus Δu has brought about a change $\Delta y = f(u_0 + \Delta u) - f(u_0)$ in y.

It is apparent in Figure 2 that the increment Δx has also occasioned a change Δy in the composite function $y = F(x)$. Now what justification is there for writing the same symbol, Δy, for the change Δf in f and also for the change ΔF in F? Well, we can see in two ways that $\Delta f = \Delta F$. Graphically, the two horizontal lines through y_0 and $y_0 + \Delta y$ on the vertical axis that measure Δf proceed parallel to each other to the right to also measure ΔF. Also, using analytic symbols, we write:

$$\Delta f = f(u_0 + \Delta u) - f(u_0)$$
$$= f[g(x_0 + \Delta x)] - f[g(x_0)].$$

But from the definition of $F(x)$,

$$\Delta F = f[g(x_0 + \Delta x)] - f[g(x_0)],$$

and equality obtains.

With the aid of the figure we can visualize three difference quotients,

$$\frac{\Delta u}{\Delta x}, \quad \frac{\Delta y}{\Delta u}, \quad \text{and} \quad \frac{\Delta F}{\Delta x} = \frac{\Delta y}{\Delta x}.$$

Working with the hint that $\Delta y/\Delta x$ can be expressed as some simple algebraic combination of $\Delta y/\Delta u$ and $\Delta u/\Delta x$, the student should experience little difficulty in seeing that it is the product of these two which gives the desired identity:

$$\frac{\Delta y}{\Delta x} = \frac{\Delta y}{\Delta u} \cdot \frac{\Delta u}{\Delta x}.$$

The approach here is different from, in fact the reverse of, the usual one. Usually upon starting the proof one writes

$$\frac{\Delta y}{\Delta x} = \frac{\Delta y}{\Delta u} \cdot \frac{\Delta u}{\Delta x},$$

and the student is encouraged to think that the right member here is obtained by a decomposition of the left member, after multiplying the left member by one, renamed $\Delta u/\Delta u$, $\Delta u \neq 0$. The graphical approach is the reverse in the sense that we first give meaning to the symbols $\Delta y/\Delta x$, $\Delta y/\Delta u$, and $\Delta u/\Delta x$, and then ask how they are related. Thus we compose, under multiplication, $\Delta y/\Delta u$ and $\Delta u/\Delta x$ to obtain $\Delta y/\Delta x$.

From the assumption that f and g are differentiable functions it follows that

$$\lim_{\Delta x \to 0} \frac{\Delta u}{\Delta x} = g'(x_0)$$

and

$$\lim_{\Delta u \to 0} \frac{\Delta y}{\Delta u} = f'(u_0).$$

The figure also allows us to see that $\Delta u \to 0$ as $\Delta x \to 0$, for the change in the function $u = g(x)$ is apparently small for a small change in x. Taking limits on both sides of the identity

$$\frac{\Delta y}{\Delta x} = \frac{\Delta y}{\Delta u} \cdot \frac{\Delta u}{\Delta x}$$

gives the result

$$F'(x_0) = f'(u_0) \cdot g'(x_0).$$

Since the point x_0 was an arbitrary point, we can drop the subscript 0 and obtain the chain rule in this form:

$$F'(x) = f'(u) \cdot g'(x),$$

or

$$\frac{dy}{dx} = \frac{dy}{du} \cdot \frac{du}{dx}.$$

The utility of a pictorial representation similar to Figure 2 does not end here, however. It can be further used to explain the difficulty with Δu being zero when Δx is not zero. For example, if $u = g(x) = k$ is a constant function then the graph of $g(x)$ becomes the horizontal line $u = k$, and $\Delta u = g(x_0 + \Delta x) - g(x_0) = k - k = 0$, for all $\Delta x \neq 0$. Under these conditions no change is induced in $y = F(x) = f[g(x)]$ by a change Δx in x_0, since

$$\Delta f = F(x_0 + \Delta x) - F(x_0)$$
$$= f[g(x_0 + \Delta x)] - f[g(x_0)]$$
$$= f[k] - f[k]$$
$$= 0.$$

But if $\Delta F/\Delta x = 0$ for all $\Delta x \neq 0$, then $F'(x_0) = 0$. Since $f'(u_0)$ is assumed to exist, we have

$$F'(x_0) = 0 = f'(u_0) \cdot 0,$$

and the chain rule appears to hold true in this case.

Now suppose that $u = g(x)$ is not a constant function, but does drop back toward the horizontal axis [thinking of

$g(x)$ as pictured in Fig. 2] and assumes again the value $u_0 = g(x_0)$ at some point $x_0 + \Delta x'$, $\Delta x' \neq 0$. Then $\Delta u = g(x_0 + \Delta x') - g(x_0) = u_0 - u_0 = 0$. Such behavior is by no means unusual for functions, and the student can see clearly how it can happen that $\Delta u = 0$ when $\Delta x \neq 0$. Now it can be pointed out that although this can happen, all but a very few of the non-constant functions that are dealt with in elementary calculus are such that when $|\Delta x|$ is small Δu will be different from zero. For it to be otherwise, $u = g(x)$ would have to assume the value u_0 infinitely many times in every neighborhood of x_0, and a function that does this is not often encountered in the beginning calculus course.

Bibliography: *Differentiation— Pedagogy*

Viertel, William K. Letter to the editor. 55 (March 1962): 164.
  Some remarks on clarifying the formula for differentiating the power of a function.

Berndt, K. L. Letter to the editor. 55 (October 1962): 451, 492.
  A pedagogical suggestion concerning the chain rule.

Viertel, William K. Some Really New "New Mathematics." 58 (October 1965): 501.
  Cites an ingenious "proof" for differentiating $1/x$.

Kregg, James F. Formulation of the Differentiation Algorithm. 64 (January 1971): 73–77.
  Pedagogical suggestions concerning the differentiation of $y = ax^n$.

Degree of accuracy of square roots

C. STANLEY OGILVY, *Hamilton College,
Clinton, New York.*

The author reminds us that it doesn't pay to generalize hastily

AN ODD MISCONCEPTION has crept into some textbooks and thence into the classroom, namely, that the process of extracting arithmetic square root yields a result which has only half as many significant figures as the original number. A specific example from a current book will perhaps recall for the reader similar statements made by his instructors in years gone by. The square root of 27.36 is calculated in the usual fashion to be 5.2, whereupon the author comments that "we are not justified in obtaining an answer beyond 5.2 as the nearest square root." We choose this sample because it comes not from some forgotten schoolbook of the nineteenth century but from the "arithmetic review" section of a college textbook published in 1952 by a highly reputable company which takes pride in the general excellence and accuracy of its texts.

The fact is that always three and essentially four figures in the answer are justified in an example like the above. It is not hard to see where the mistaken notion comes from. The arithmetic is displayed as follows:

$$
\begin{array}{r}
5 \cdot 2 \\
\sqrt{27 : 36} \\
25 \\
\hline
\end{array}
$$

$$
\begin{array}{r|rr}
102 & 2 & 36 \\
& 2 & 04 \\
\hline
\end{array}
$$

"It is obvious" that the next figure in the answer depends on the next two (missing) figures in the given number, and hence that we cannot safely supply a pair of zeros after the 27.36. This is just another example of an "obvious" thing that isn't so. In long division, it is so; and the mind jumps over-zealously to promote a false analogy. Let us investigate the general situation and then return to a further consideration of this example.

The method of approximation by differentials is helpful. If $y = x^{1/2}$, $dy = \frac{1}{2}x^{-1/2}dx$. Let N be the given number whose square root is to be taken. Let x be the same number shorn of the decimal point and any terminal zeros. (For instance, if $N = 27.36$ or 273600, then $x = 2736$.) Take first the case where the decimal point has been moved an even number of places: $x = 100^k N$, k a positive or negative integer or zero. Then $y = \sqrt{x} = 10^k\sqrt{N}$, $dx \leq \frac{1}{2}$, and $dy \leq 1/(4\sqrt{x})$. We can tabulate dy as follows:

If $10^0 \leq x < 10^1$, $\sqrt{x} \geq 1$, $dy \leq .25$ < 1.0

(first significant figure)

If $10^1 \leq x < 10^2$, $\sqrt{x} > 3$, dy $< .1$

(second significant figure)

If $10^2 \leq x < 10^3$, $\sqrt{x} \geq 10$, $dy \leq .0025 < .01$

(third significant figure)

If $10^3 \leq x < 10^4$, $\sqrt{x} > 30$, dy $< .001$

(fourth significant figure)

—and so on.

It is evident that these inequalities could be considerably sharpened; but it is enough for our purposes to note that in no case can the square root of a number having m significant figures contain an error of more than unity in the m'th significant figure.

If the decimal point has to be moved an

odd number of places in the transition from N to x, we can make the substitution $z = 10x$. Then $dz = 10dx$, and dy is increased by a factor of $\sqrt{10}$. But of course this factor is divided out again in the final return from $\sqrt{10x}$ to $\sqrt{x}$, and we are left with the same result as before.

Returning now to the original textbook example, let us assume that 27.36 has been rounded off from some longer decimal. Then the "exact" value must lie between 27.355 and 27.365. But

$$\sqrt{27.35\ 50\ 00\ 00} = 5.2302, \text{ and}$$

$$\sqrt{27.36\ 50\ 00\ 00} = 5.2312.$$

Thus we see that regardless of what the exact value was, the round-off does not affect the square root until the fourth significant figure. Furthermore, if we calculate $\sqrt{27.36\ 00\ 00\ 00}$ from scratch (not interpolating between the other two), we obtain, of course, 5.2307, the mid-value. Rounding off to 5.231 introduces an error of at most one in the fourth figure, as predicted by the theory.

In general, then, except for a possible error of ± 1 in the last significant figure, one obtains a square root accurate not to half as many but to the same number of significant figures as were given in the original number.

For an extended treatment of the maximum error to be expected in extracting not only square roots but also cube, fourth, fifth, and sixth roots, the interested reader is referred to Section 14 of *Introduction to Numerical Analysis*, F. B. Hildebrand (McGraw-Hill Book Co., Inc., 1956).

STRAIGHT-LINE TUNNELS THROUGH THE EARTH

By **WILLIAM C. STRETTON**

Lyons Township High School and Junior College
La Grange, Illinois

SUPPOSE there is a straight-line tunnel through the earth connecting any two points on the earth's surface, say P and Q. Suppose, further, that it is possible to eliminate all friction and air resistance inside the tunnel. Then an object released at P would be accelerated by the earth's gravity until it got to M, the midpoint of PQ as shown in Figure 1. Once at M, its kinetic energy would be just sufficient to

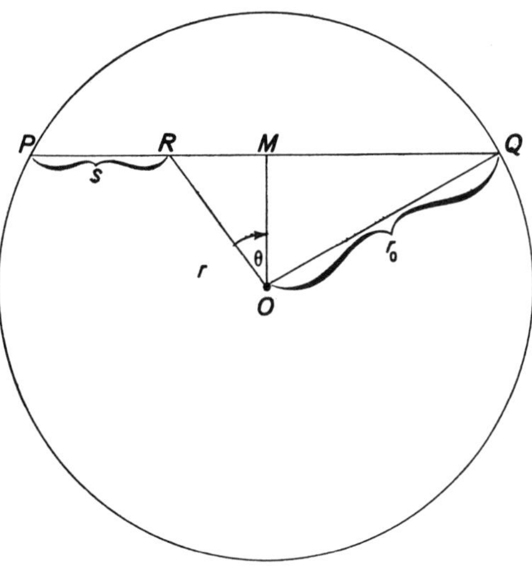

FIGURE 1

overcome gravity and carry it through to Q, the other end of the tunnel. It is the purpose of this paper to derive velocity and distance relations for the trip through the tunnel and in particular to show that the time required for the trip is constant, independent of the locations of P and Q.

Now it is known that the acceleration due to gravity at a point beneath the surface of the earth is directly proportional to the distance from the point to the center—that is, $\alpha = kr$ where r represents the distance from the center and α is the acceleration. It is easy to determine k, since when $r = r_0$, the radius of the earth, then $\alpha = g$. Hence $k = \dfrac{g}{r_0}$.

Suppose the object at any time t is at R, and denote the distance traveled from P by s. The component of acceleration in the direction of PQ is easily seen from the figure to be $\alpha \sin \theta$—that is, $\dfrac{dv}{dt} = \dfrac{gr}{r_0} \sin \theta$. Now if we let $PM = b$, then it is also easily seen from the figure that

$$\sin \theta = \frac{b - s}{r}.$$

Hence

$$\frac{dv}{dt} = \frac{g}{r_0}(b - s).$$

Now, by the chain rule,

$$\frac{dv}{dt} = \frac{dv}{ds} \cdot \frac{ds}{dt} = v\frac{dv}{ds}.$$

Therefore we have

$$v\frac{dv}{ds} = \frac{g}{r_0}(b - s).$$

In this simple differential equation, the variables are easily separated, and integration of both sides yields

$$\frac{v^2}{2} = \frac{g}{r_0}\left(bs - \frac{s^2}{2}\right) + C_1.$$

The constant C_1 can be determined from the fact that when $s = 0$, $v = 0$; so it follows that $C_1 = 0$.

Let us write the velocity relation as

$$v^2 = \frac{g}{r_0}(2bs - s^2).$$

Then

$$v = \sqrt{\frac{g}{r_0}}(2bs - s^2)^{\frac{1}{2}}.$$

Using the fact that $v = \dfrac{ds}{dt}$ we have another differential equation, namely

$$\frac{ds}{dt} = \sqrt{\frac{g}{r_0}}(2bs - s^2)^{\frac{1}{2}}.$$

The variables are easily separated, and we get

$$\frac{ds}{\sqrt{2bs - s^2}} = \sqrt{\frac{g}{r_0}}\, dt.$$

In order to integrate this, we notice that

$$2bs - s^2 = b^2 - (b^2 - 2bs + s^2) = b^2 - (s - b)^2.$$

We then have

$$\frac{ds}{\sqrt{b^2 - (s - b)^2}} = \sqrt{\frac{g}{r_0}}\, dt.$$

We can now integrate the left side, using a standard integral table, and we get

$$\operatorname{arc\,sin} \frac{s - b}{b} = \sqrt{\frac{g}{r_0}}\, t + C_2.$$

C_2 can now be determined from the fact that when $t = 0$, $s = 0$; so we have

$$\operatorname{arc\,sin}(-1) = C_2.$$

Hence

$$C_2 = -\frac{\pi}{2}.$$

Therefore the equation relating s and t is

$$\operatorname{arc\,sin} \frac{s - b}{b} = t\sqrt{\frac{g}{r_0}} - \frac{\pi}{2}.$$

Now we can calculate the time required to reach M; and, by doubling, the total time for the trip from P to Q. For when the object is at M, then $s = b$ and we have

$$\operatorname{arc\,sin} 0 = t\sqrt{\frac{g}{r_0}} - \frac{\pi}{2}.$$

Hence

$$t = \frac{\pi}{2}\sqrt{\frac{r_0}{g}}.$$

So the time required to go from P to Q is

$$\pi\sqrt{\frac{r_0}{g}}.$$

If we take $g = \dfrac{32 \text{ ft}}{\sec^2}$ and $r_0 = 4{,}000$ miles, then the time for the trip turns out to be about 42 minutes. Since P and Q were any points on the surface of the earth, it follows that the time required to travel any straight line tunnel is a constant 42 minutes.

Some interesting sidelights follow from the above calculations. Notice the similarity between the time equation for the complete trip,

$$t = \pi\sqrt{\frac{r_0}{g}},$$

and the familiar formula for the period of a simple pendulum,

$$T = \pi\sqrt{\frac{l}{g}}$$

where l is the length of the pendulum. In other words, the time for a tunnel trip is exactly the period of a pendulum whose length is the radius of the earth!

Also, if our tunnel passes through the center of the earth, then $b = r_0$, and our velocity equation becomes

$$V = \sqrt{\frac{g}{r_0}}(2r_0 s - s^2)^{\frac{1}{2}}.$$

The speed at the center is then found by setting $s = r_0$. We get

$$V = \sqrt{\frac{g}{r_0}}(2r_0{}^2 - r_0{}^2)^{\frac{1}{2}} = \sqrt{gr_0}.$$

Since the velocity of escape from the surface of the earth is known to be $\sqrt{2gr_0}$, we see that the ratio of the escape velocity to the maximum tunnel velocity is $\sqrt{2}$.

Bibliography: *Applications of Differentiation*

Stover, Donald W. Projectiles. 57 (May 1964): 317–22.

Derivation of projectile equations by noncalculus means; includes suggested apparatus for classroom demonstration.

Hirsch, A. Adler. An Optical Method for Demonstrating the Point of Inflection of Curves. 59 (March 1966): 250.

Remarks on a method for determining the points of inflection by images in a doubly reflecting mirror.

Mandelbaum, Joseph, and Albert Schild. An Interesting Relationship among the Roots of a Cubic Equation. 63 (May 1970): 393–94.

Disproves a student-held belief that the tangent line to a curve at a point whose abscissa is midway between two of its real zeros is always horizontal.

Cummins, Kenneth. Mathematics "In Statu Nascendi." 63 (November 1970): 567–70.

Suggestions for teaching slope and curvature using student creativity as opposed to formal deductive presentations.

Calculus Versus Geometry

By CLAIRE FISHER ADLER
New York University
New York City

SOMETIME ago this problem was given on an examination for teaching positions in the high schools of New York City.

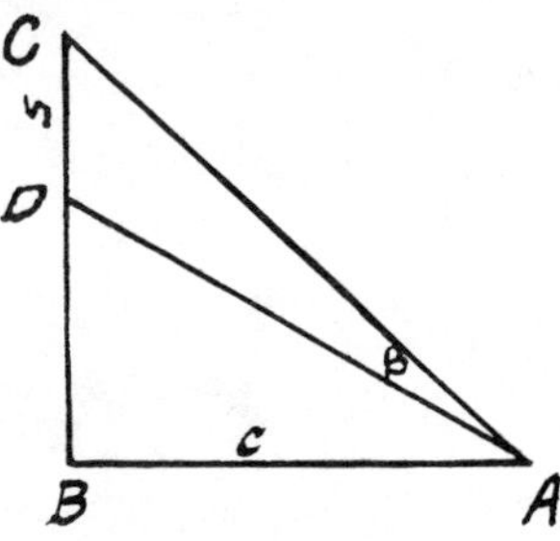

Given a right triangle ABC with CB a constant, CD a fixed length s, $\beta = \angle CAD$, $AB = c$. To show that when β is a maximum $s = 2c \tan \beta$.

It seemed that the main difficulty in handling the problem was not in finding a solution but in finding the required geometric one for a problem involving the maximum value of a variable. Such treatments are usually inserted in an isolated section in textbooks on Plane Euclidean Geometry[1] and due to the limited time at the disposal of teachers of these courses this subject material is seldom touched; consequently, little is known to students in these courses about the treatment of such problems. Moreover in many cases these problems are handled somewhat laboriously and awkwardly by the geometric or purely analytic method whereas a calculus solution is extremely simple. For this reason it occurred to the writer that it would be an interesting novelty to compare the classic geometric solutions of some of these problems with the more elegant and powerful methods of the calculus. Two such solutions of each of the following four problems have been given,

the first three of which are found treated geometrically in the texts before referred to[2] and the fourth is the problem mentioned at the beginning of this article.

PROBLEM 1

Of all triangles having two given sides, the area of that in which these sides include a right angle is the maximum.

Geometry

Let the triangles ABC and EBC have the sides AB and BC equal to EB and BC respectively; and let the angle ABC be a right angle.

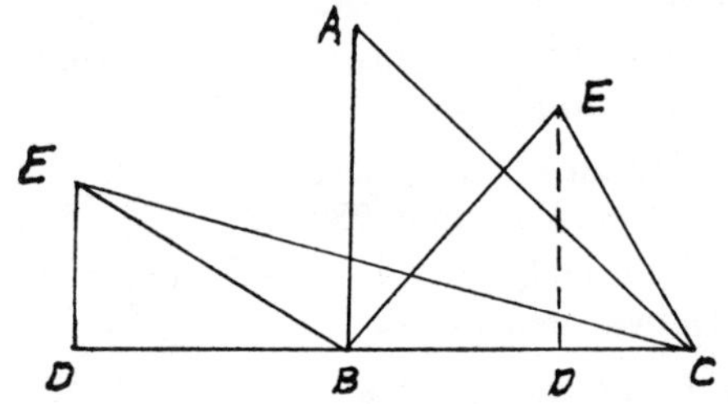

To prove that $\triangle ABC > \triangle EBC$.

Proof: From E draw the altitude ED. The triangles ABC and EBC having the same base, $= BC$, are to each other as their altitudes, AB and ED.

Now $EB > ED$.

But $EB = AB$.

$\therefore AB > ED$.

$\therefore \triangle ABC > \triangle EBC$.

Calculus

Given triangle ABC with given sides $AC = b$ and $BC = a$ and these sides including $\angle ACB = \theta$.

Then area $S = \frac{1}{2}ab \sin \theta$.

[1] *Plane and Solid Geometry*, Wentworth, Revised Edition, Page 234.

[2] *Plane and Solid Geometry*, Ford and Ammerman (1919) Appendix, Page 206.

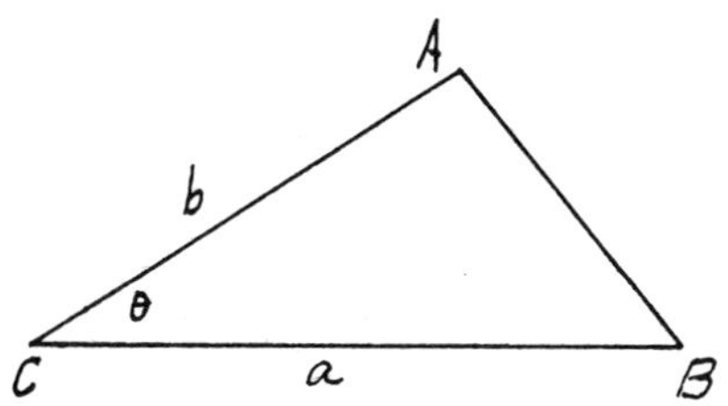

For a maximum $d_\theta S = \frac{1}{2}ab \cos \theta = 0$.

Therefore $\theta = 90°$.

Problem 2
Geometry

Of all isoperimetric triangles having the same base the area of the isoscles triangle is the maximum.

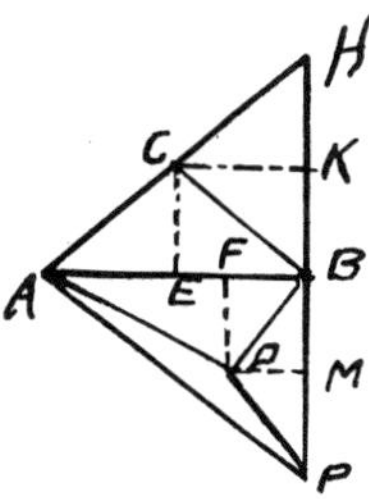

Let the triangles ACB and ADB have equal perimeters and let AC and BC be equal and let AD and DB be unequal.

To prove that $\triangle ABC > \triangle ADB$.

Proof Produce AC to H, making $CH = AC$; and draw HB. Produce HB, take DP equal to DB and draw AP. Draw CE and $DF \perp AB$, and CK and $DM \| AB$.
The angle ABH is a right $\angle$ for it may be inscribed in the semicircle whose centre is C and radius CA.
ADP is not a straight line, for then the $\angle s \, DBA$ and DAB would be equal, being complements of the equal $\angle s \, DBM$ and DPM respectively and DA and DB would be equal which is contrary to the hypothesis. Hence

$$AP < AD + DP$$
$$< AD + DB$$
$$< AC + CB$$
$$< AC + CH$$
$$< AH.$$
$$\therefore BH > BP.$$
$$\therefore CE(= \tfrac{1}{2}BH) > DF(= \tfrac{1}{2}BP).$$
$$\therefore \triangle ACB > \triangle ADB.$$

Calculus

Given ABC with $AB = a$ and, a given perimeter $= 2k$

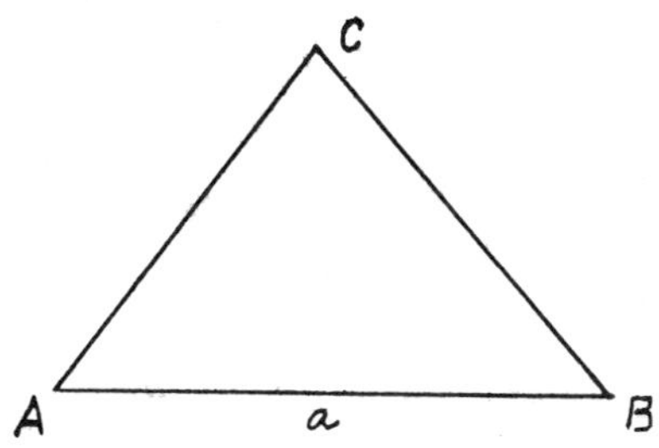

Let $AC = x$.

Then $BC = 2k - (a + x)$.

Now area $S = \overline{K}\sqrt{(k-x)(a+x-k)}$.

where $\overline{K} = \sqrt{k(k-a)}$

and
$$d_x S = \overline{K}\frac{\{(k-x)+(a+x-k)(-1)\}}{2S} = 0.$$

$$\therefore \quad k - x = a + x - k.$$

or
$$2k - (a+x) = x.$$

$$\therefore \quad BC = AC.$$

Problem 3
Geometry

Of all polygons with sides all given but one, the one with a maximum area can be inscribed in a semicircle which has the undetermined side for its diameter.

This problem reduces to Problem 1 for a polygon with three sides. For four sides the area will be a function of more than one variable. The calculus discussion will be limited to this case.

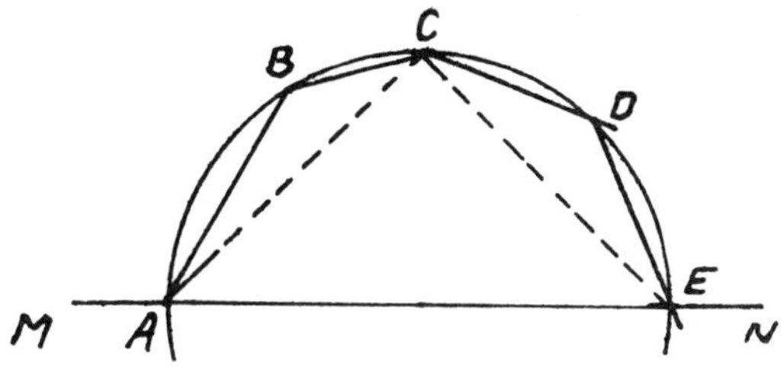

Let $ABCDE$ be the maximum area of polygons with sides AB, BC, CD, DE and the extremities A and E on the straight line MN.

To prove that $ABCDE$ can be inscribed in a semicircle.

Proof From any vertex as C, draw CA and CE. The $\triangle ACE$ must be the maximum of all triangles having the sides CA and CE and the third side on MN: otherwise, by increasing or diminishing the $\angle ACE$, keeping the lengths of the sides CA and CE unchanged, but sliding the extremities A and E along the line MN, we could increase the $\triangle ACE$ while the rest of the polygon would remain unchanged; and therefore increase the polygon. But this is contrary to the hypothesis that the polygon is the maximum.

Hence, the $\triangle ACE$ is the maximum of $\triangle$ that have the sides CA and CE.

$\therefore ACE$ is a right angle.

C lies on the semicircumference; that is, the maximum polygon can be inscribed in a semicircle having the undermined side for a diameter.

Calculus

Let $ABCD$ be a polygon with sides $AB = a$, $BC = b$, $CD = c$ where a, b, c are fixed positive quantities.

To prove that $ABCD$ can be inscribed in a semicircle when $ABCD$ is the maximum polygon.

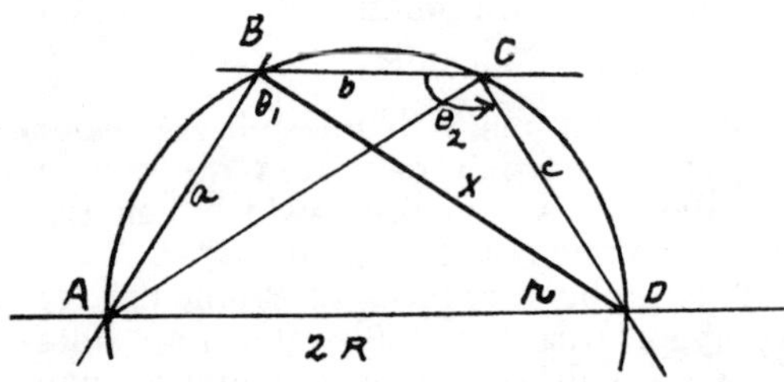

Proof Let $AD = 2R$, a positive quantity; $\angle ABD = \theta_1$, $\angle BCD = \theta_2$, $\angle BDA = r$.

Then $S = \text{area} = \frac{1}{2}(ax \sin \theta_1 + bc \sin \theta_2)$ where $x = \sqrt{b^2 + c^2 - 2bc \cos \theta_2}$.

Hence S is a function of two variables. For a maximum,*

(1) $\dfrac{\partial S}{\partial \theta_1} = 0$ $\dfrac{\partial S}{\partial \theta_2} = 0$

also

(2) $\left\{\dfrac{\partial^2 S}{\partial \theta_1^2} \dfrac{\partial^2 S}{\partial \theta_2^2} - \left(\dfrac{\partial^2 S}{\partial \theta_1 \partial \theta_2}\right)^2\right\} > 0$

---

* *Granville, Smith and Longley.* Elements of Differential and Integral Calculus, p. 449.

and $\dfrac{\partial^2 S}{\partial \theta_i^2} < 0$ $i = 1, 2.$

Now $\dfrac{\partial S}{\partial \theta_1} = \underset{x \neq 0}{\tfrac{1}{2}ax} \cos \theta_1 = 0$ $\therefore \theta_1 = 90°.$

$\therefore \tan \angle BAD = \dfrac{x}{a}.$

$\dfrac{\partial S}{\partial \theta_2} = \dfrac{bc}{2}\left[\dfrac{a \sin \theta_1 \sin \theta_2}{x} + \cos \theta_2\right] = 0.$

But $\sin \theta_1 = 1$ $\therefore \dfrac{a \sin \theta_2}{x} + \cos \theta_2 = 0.$

$\therefore \tan \theta_2 = -\dfrac{x}{a}.$

Therefore θ_2 is the supplement of $\angle BAD = 90 - r.$

$\therefore \theta_2 = 90 + r.$

But since ABD is a right triangle a circle may be drawn through ABD.

$\therefore$ arc $\overset{\frown}{BAD} = 180° + 2r$; hence arc $\overset{\frown}{BD}$ is the locus of vertices of all angles equal to $90 + r$ having BD as a subtended chord. Hence C lies on this arc and the polygon $ABCD$ is inscribed in the circle having AD for a diameter.

Also

$$\left.\dfrac{\partial^2 S}{\partial \theta_1^2}\right]_{\substack{\theta_1 = 90° \\ \theta_2 = \tan^{-1} - x/a}} = -\tfrac{1}{2}\,ax$$

But since a and x are both positive $ax < 0.$

$$\therefore \left.\dfrac{\partial^2 S}{\partial \theta_1^2}\right]_{\substack{\theta_1 = 90° \\ \theta_2 = tan^{-1} - x/a}} < 0.$$

$$\left.\dfrac{\partial^2 S}{\partial \theta_2^2}\right]_{\substack{\theta_1 = 90° \\ \theta_2 = \tan^{-1} - (x/a)}} = -\dfrac{bc}{2x}\left[\dfrac{(a^2 + x^2)^{3/2} + abc}{a^2 + x^2}\right]$$

$$= -\dfrac{bc}{2x}\left[\dfrac{(2R)^{3/2} + abc}{4R^2}\right] < 0.$$

Also

$$\left[\dfrac{\partial^2 S}{\partial \theta_1^2} \dfrac{\partial^2 S}{\partial \theta_2^2} - \left(\dfrac{\partial^2 S}{\partial \theta_1 \partial \theta_2}\right)^2\right]$$

$$= abc\left(\dfrac{(2R)^{3/2} + abc}{16R^2}\right) > 0.$$

$\therefore ABCD$ is the maximum polygon.

Problem 4
Geometry

Given a right triangle ABC with CB a constant, CD a fixed length s, $\beta = CAD$,

$AB = c$. To show that when β is a maximum $s = 2c \tan \beta$.

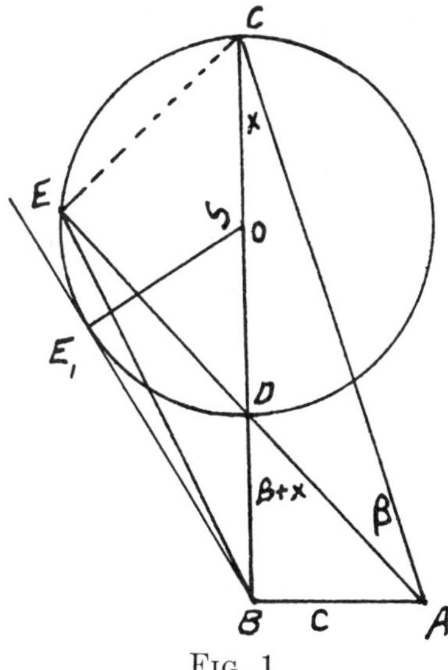

FIG. 1

Draw circle with center 0 and radius $s/2$
Construction Extend line AD until it cuts this circle at E. Connect B and E.

Let $\qquad \angle BCA = x$.

Then $\qquad \angle BDA = \beta + x$.

Now

$$\triangle EDC \backsim \triangle BDA \qquad (3 \angle =)$$

$$\therefore \quad \frac{ED}{DB} = \frac{DC}{DA}$$

also

$$\angle EDB = \angle CDA .$$

$$\therefore \quad \triangle EDB \backsim \triangle CDA$$

$$(\angle = \text{including sides} \therefore)$$

$$\therefore \quad \angle BED = x \quad \text{and} \quad \angle EBD = \beta .$$

Now when E is at D $\angle \beta = 0$. As E moves from D to C $\angle \beta$ varies. When E is at C, β is again 0. Hence for some position of E on the semicircle β is a maximum. Consider the tangent line to the circle i.e. BE_1.

The $\angle E_1BC > \angle EBC$ since the secant line EB falls to the right of the tangent line E_1B.

Hence $\angle E_1BC$ is a maximum.

To prove When BE is tangent to the circle $S = 2c \tan \beta$.

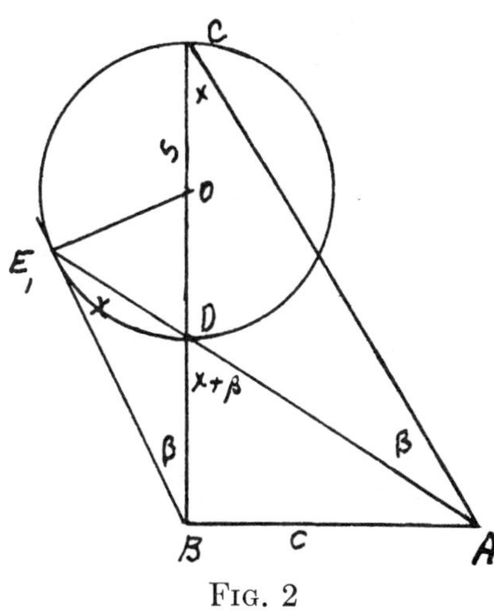

FIG. 2

$$\angle DE_1O = 90 - x = \angle E_1DO = x + \beta .$$

Hence

(1) $\quad 90 - x = x + \beta$; or $90 - \beta = 2x$ or

$\qquad 90 - (\beta + x) = x = \angle AE_1B = \angle DAB$.

$\qquad \therefore$

(2) $\quad \triangle E_1BA$ is isosceles

$\qquad \therefore$

(3) $\quad E_1B = BA = c$.

But

(4) $\quad \tan \beta = \dfrac{E_1O}{E_1B} = \dfrac{s/2}{c}$.

$\qquad \therefore$

(5) $\quad s = 2c \tan \beta$.

Calculus*

Using Fig. 2. as above and letting $DB = h$.

(1) $\quad \dfrac{s}{\sin \beta} = \dfrac{AD}{\sin x} = \dfrac{h}{\sin x \cos (x + \beta)}$

(2) $\therefore \quad s \sin x \cos (\beta + x) - h \sin \beta = 0$.

Differentiating and solving for $d\beta/dx$

(3)

$$\frac{d\beta}{dx} = s \left[\frac{\cos (\beta + x) \cos x - \sin (\beta + x) \sin x}{s(\sin x) \sin (\beta + x) + h \cos \beta} \right].$$

* This problem is the classic picture or tapestry problem found in many standard calculus text books. See Problem 17. P. 114 in reference given in Problem 3.

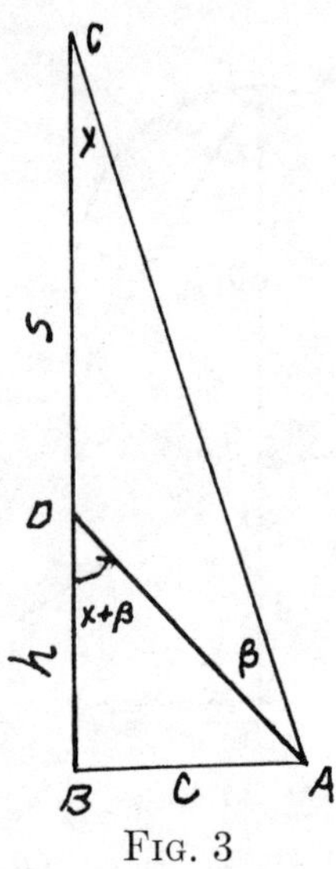

Fig. 3

Hence if

$$\frac{d\beta}{dx} = 0$$

$$\cos(\beta+x)\cos x - \sin(\beta+x)\sin x = 0$$

$$\therefore \quad \cos(\beta+2x) = 0.$$

(4) $\quad \therefore \quad \beta+2x = 90° \quad \therefore \quad \sin 2x = \cos \beta.$

But

(5) $\quad \dfrac{s}{\sin \beta} = \dfrac{c}{\sin x \sin(x+\beta)} = \dfrac{c}{\sin x \cos x}$

$$= \frac{2c}{\sin 2x} = \frac{2c}{\cos \beta}.$$

(6) $\quad \therefore \quad s = 2c \tan \beta.$

A comparison of these four problems reveals the superiority of the methods of the calculus in the second and fourth problems only. In the first problem geometry was declared more satisfactory by students in a course in general mathematics where very elementary ideas of calculus had been introduced. The superiority of the geometric solution is demonstrated in problem three where even a comparatively simple application of the calculus leads to rather lengthy computations. However in the fourth problem the efficiency of the calculus is clearly demonstrated. It has led to a neat and clear solution, without the introduction of any special device, lengthy explanations of construction, discussion or detailed proof. It is surely a problem admirably adapted to delight the ardent adherents to the methods of the calculus!

COMPUTER-EXTENDED INSTRUCTION: An Example

By William S. Dorn, University of Denver, Denver, Colorado

Introduction

IN THIS article we will discuss the computer and its use as a laboratory in conjunction with a traditional mathematics course. This type of computer use has a direct analogy with the use of a physics laboratory as it applies to a physics recitation class. We will make this analogy sharper in the next paragraph. Before doing so, however, we point out that the mathematics used in this article should be within the reach of most 8th or 9th grade students. That is, it uses only elementary algebra including some inequalities. However, the problems are interesting and difficult enough to challenge 12th grade students and, in fact, any pre-calculus student.

In order to put what follows in its proper perspective, we review briefly the rationale behind any science laboratory. Such a laboratory generally is used (*a*) to *motivate* the student to delve further into the science, (*b*) to develop the student's *intuition* regarding certain aspects of the science, and (*c*) to allow the student to make certain "educated guesses" or *conjectures* regarding some aspect or aspects of the science. For example, suppose we place a physics student in a laboratory with a meter stick, some blocks with different weights, and a fulcrum. We then ask him to balance the meter stick on the fulcrum using the blocks in pairs. We can expect that the student will soon notice that the heavier blocks should be placed closer to the fulcrum than the lighter blocks. Thus we develop his *intuition* regarding mechanical advantage.

If the student is clever and/or perseverent he may discover that if he replaces a block with another block having twice the weight of the first, then he must halve the distance from the fulcrum. Thus he can *conjecture* that the product of the weight of a block and the block's distance from the fulcrum is a constant.

Of course, the student cannot verify this conjecture until he has studied some mechanics in his physics recitation class. Thus the use of this particular laboratory has *motivated* the student to study physics in order to test his own theories about mechanical advantage.

In this paper we are concerned with mathematics, not physics. We will discuss some simple maximization and minimization problems and use a computer to perform some experiments with them. The objectives of these experiments are:

1. To develop some insight into the mathematical concepts of maximum and minimum
2. To allow the student to "guess" at some theorems regarding when a maximum or minimum occurs
3. To motivate the student to study the calculus.[1]

---

1. While algebra will carry the student through most of the material here, including a proof of one theorem (under "Verification of a Conjecture"), he will find that to go much further with maximization problems he will need the calculus at his disposal.

A thread that runs through all of this is a description of how a mathematician "does" mathematics. That is, this paper is a simple example of how a mathematician develops his intuition and insight to the point where he can state what he thinks may be a theorem. This is followed by an attempt to prove the "theorem." In this way we answer, at least partially, the beginning student's perennial question: "How did anyone ever think up those theorems in my textbook anyway?"

All of the programs in this paper are written in BASIC. There are three reasons for the choice of this particular computer language. First, it is easily learned even by elementary school children. Secondly, it is "conversational." That is, one can easily and quickly alter a BASIC program. We will need this facility because most of our nine programs will build upon their predecessors. In some cases the program statements (or sentences) themselves depend on the arithmetic results of the preceding program. Without a language having a conversational capability the development here would be awkward, to say the least. Finally, we use BASIC because it is widely available. The computers on which the BASIC language is available are legion. Only FORTRAN outranks it in popularity in the scientific community.

In addition to the mathematical lessons noted above there are some programming lessons to be learned in our development. The reader unfamiliar with BASIC and/or maximization problems might be well advised to read the predecessor to this article [3][2] or the extremely elegant NCTM booklet [1]. An excellent tutorial on all of the facilities available in BASIC can be found in the first part of the Kemeny and Kurtz textbook [4].

In the predecessor to this article [3] we developed a simple loop for handling a maximization problem. The present article starts with the last program of [3] and

develops the concept of "nested loops." Moreover, in this article we will discuss the value of performing some preliminary mathematical analysis in order to reduce the amount of computer time consumed and to reduce the amount of data that is printed. Perhaps even more important is the discussion of how and why we should have the computer do some bookkeeping work for us. This latter point is, of course, closely tied to reducing the amount of printing to be done.

With this as background we now proceed to discuss some particular experiments. We will use the results of these experiments to make some conjectures. In one case we will actually prove that our conjecture is valid. Along the way we will discuss at length just what constitutes a mathematical proof and what part a computer can and cannot play in a proof.

The Fencing Problem

Consider the following problem. Suppose we have a given total length of fencing, F. This fencing is in sections each of length S. We wish to form a rectangular yard that is bounded on one side by a stone wall which is longer than F (see fig. 1). How should we use the fencing to make the area of the yard as large as possible?

In a previous paper [3] we addressed this problem by writing a sequence of nine computer programs, each one slightly more sophisticated than its predecessor. The ninth and last program in that paper is shown below together with its output (see program 1).

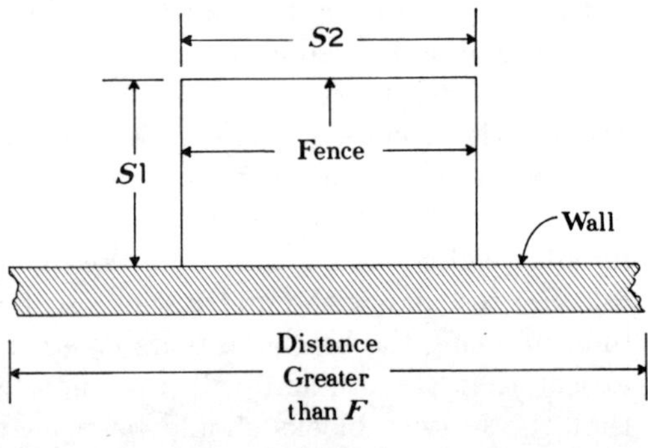

FIGURE 1

---

2. Numbers in square brackets [ ] refer to the bibliography at the end of the article.

The "SHORT SIDE" is the length of the two sides perpendicular to the wall ($S1$ in fig. 1) and the "LONG SIDE" is the one side parallel to the wall ($S2$ in fig. 1). For more details about the program see the "PROGRAM NOTES" immediately following the output· of the program.

PROGRAM 1

```
5    READ F,S
7    IF S = Ø THEN 11Ø
8    LET M = Ø
1Ø   FOR S1 = S TO F/2 STEP S
2Ø       LET S2 = F − 2*S1
3Ø       LET A = S1*S2
4Ø       IF A < = M THEN 8Ø
5Ø       LET M = A
6Ø       LET L1 = S1
7Ø       LET L2 = S2
8Ø   NEXT S1
9Ø   PRINT "TOTAL FENCING",F
91   PRINT "PIECE LENGTH IS",S
92   PRINT "LARGEST AREA IS",M
93   PRINT "SHORT SIDE IS",L1
94   PRINT "LONG SIDE IS",L2
95   PRINT
96   PRINT
97   GO TO 5
1ØØ  DATA 1ØØ,1,1ØØ,.333333,51,.25, 51,.125
11Ø  END
```

The output from a run of program 1 is as follows:

```
RUN

TOTAL FENCING    1ØØ
PIECE LENGTH IS  1
LARGEST AREA IS  125Ø
SHORT SIDE IS    25
LONG SIDE IS     5Ø

TOTAL FENCING    1ØØ
PIECE LENGTH IS  .333333
LARGEST AREA IS  125Ø.
SHORT SIDE IS    25.
LONG SIDE IS     5Ø.ØØØ1

TOTAL FENCING    51
PIECE LENGTH IS  .25
LARGEST AREA IS  325.125
SHORT SIDE IS    12.75
LONG SIDE IS     25.5

TOTAL FENCING    51
PIECE LENGTH IS  .125
LARGEST AREA IS  325.125
SHORT SIDE IS    12.75
LONG SIDE IS     25.5

OUT OF DATA IN   5
```

PROGRAM NOTES: The reader totally unfamiliar with BASIC would do well to refer to [3] where nine successive programs are created to build up to the present program. Statement 5 reads values of F (the total length of fencing available) and S (the standard lengths in which the fencing is made). If the section length S is zero, we stop (see statement 7), since $S = 0$ does not make sense. Actually we should stop if $F \leq 0$ and also if $S > F$, but we have not incorporated these features in this program.

The variable M is the maximum area computed to date, so we start with $M = 0$ (see statement 8) since at the start we have not computed any areas. There are two sides perpendicular to the wall and if each exceeds $F/2$ then the amount of fencing used exceeds F. Therefore, $S1$, the length of the side perpendicular to the wall, cannot exceed $F/2$. The FOR statement (statement 10) lets $S1$ take on the values S, $2S, \ldots, F/2$.

In the "loop" initiated by the FOR statement, we calculate $S2$ (the length of the side parallel to the wall) as $F - 2 \cdot S1$ and $A = S1 \cdot S2$ (the area). The asterisk (*) indicates multiplication in BASIC. If the area, A, does not exceed the maximum area computed to date, then we go to statement 80 which increments $S1$ by S. However, if we have computed a larger area, we set M equal to that area (statement 50) and let $L1$ and $L2$ be the corresponding dimensions which yield this new maximum area. We then increase $S1$ by S.

After $S1 = F/2$ we print (see statements 90 through 94) the total fencing, section length, maximum area possible, the lengths of the sides perpendicular to the wall (SHORT SIDE), and the length of the side parallel to the wall (LONG SIDE). We then print two blank lines and return to statement 5 to read another pair of values for F and S.

The DATA statement (100) says first let $F = 100$ and $S = 1$, then let $F = 100$ and $S = 1/3$. The third case sets $F = 51$ and $S = 1/4$. Finally we set $F = 51$ and $S = 1/8$.

It appears from the output of program 1 that the length of the one side parallel to the wall ($S2$) should be twice the length of the two sides perpendicular to the wall ($S1$), as stated in equation 1.

$$(1) \qquad S2 = 2 \cdot S1 = F/2.$$

Through our BASIC program we have *proved* that equation (1) is valid for four specific cases,

$$(a) \quad F = 100 \text{ and } S = 1,$$
$$(b) \quad F = 100 \text{ and } S = 1/3,$$
$$(c) \quad F = 51 \text{ and } S = 1/4,$$
$$\text{and} \quad (d) \quad F = 51 \text{ and } S = 1/8,$$

because we have computed the areas for all possible dimensions and have printed out the largest area. This, of course, is *proof by exhaustion* and can only be done when a finite number of possibilities exist. Also note that without a computer, even proof by exhaustion is unreasonable if the number of possibilities is very large, as it is here.

If we let $F = 51$ and $S = 1$, we find that the maximum area is 225 and that $S1 = 13$ while $S2 = 25$. Thus $S2$ is almost $2 \cdot S1$, but not quite. Other such examples can, of course, be constructed. The facts are: In case $4 \cdot S$ is a factor of F then equation (1) holds. In any case equation (1) always is approximately true. In fact, if we/ use equation (1) if $S = 1$, then rounding off to the nearest multiples of S we get the correct result. For example, with $F = 51$ it follows from equation (1) that $S1 = 1275$. By rounding off $S = 1$ to the nearest multiple of S, we get $S1 = 13$.

We have, of course, not "proved" the above statements. Their proof is beyond the scope of this paper. We observe, however, that apparently if the section length, S, is "small enough," then equation (1) holds. When $F = 51$, equation (1) holds if $S = 1/4$ or $S = 1/8$. Equation (1) does not hold if $S = 1$. One could, of course, try $S = 1/16$ or $S = 1/20$ with $F = 51$ to further verify the *conjecture* that equation (1) is valid provided S is sufficiently small.[3] A proof of this conjecture is trivial if one uses the calculus.

Even without the calculus, however, we can prove this conjecture that if the section length, S, is arbitrarily small, then equation (1) must hold. We proceed to that now.

Verification of a Conjecture

We again address the three-sided fencing problem described at the beginning of the previous section. We suppose, however, that the fencing is made of flexible wire so that we can make the section lengths anything we wish. It follows then that: *The one side parallel to the wall should be twice as long as the two sides perpendicular to the wall,* or

$$S2 = 2 \cdot S1 = F/2.$$

Let L be the length of the two sides perpendicular to the wall which makes the area A a maximum. We will show that this means that $L = F/4$.

If L is the length of the sides perpendicular to the wall then the length of the side parallel to the wall is $F - 2L$. The formula for maximum area is given below.

$$(2) \qquad A = LF - 2L^2.$$

We reiterate that this is the maximum area *by assumption*. Note that at this point the "value" of L is unknown. From the fact that the area A in equation (2) is a maximum we will deduce the value of L.

If we change L ever so slightly, then the new area must be smaller than A. (Remember A is the maximum area.) Let us compute the area for the case where the sides perpendicular to the wall differ from L. To this end we let the sides perpendicular to the wall have a length $L + e$ where e may be any positive or negative number (not zero). The side parallel to the wall then has a length $F - 2(L + e)$. The area, B, can be found by the formula

$$B = (L + e)[F - 2(L + e)]$$

which is restated in equation (3).

$$(3) \qquad B = LF - 2L^2 + e(F - 4L - 2e).$$

Now since by assumption A is larger than B, it must be that $A - B$ is positive. Equations (2) and (3), then, lead us to

$$(4) \qquad A - B = e(2e + 4L - F) > 0.$$

Now two cases arise: the value of e is either greater than zero or less than zero. The inequality (4) must hold in *both* cases. We consider first $e = e_1 > 0$, then from (4) it must be that

$$2e_1 + 4L - F > 0.$$

---

3. If we use $S = 1/10$ we will find that equation (1) is only approximately correct, but that the deviation from equation (1) will be much smaller than it was when $S = 1$.

Thus

$$4L - F > -2e_1.$$

Since $e_1 > 0$,

$$(5) \qquad 4L - F > -|2e_1|.$$

Next we consider the second case, where $e = e_2 < 0$. If (4) is to be satisfied it must be that

$$2e_2 + 4L - F < 0$$

or

$$4L - F < -2e_2.$$

Since $e_2 < 0$, it follows that

$$(6) \qquad 4L - F < |2e_2|.$$

Combining (5) and (6),

$$(7) \qquad -|2e_1| < 4L - F < |2e_2|$$

for *all* values of $|e_1|$ and $|e_2|$.

The only[4] way both inequalities can be satisfied for all "very small" values of $|e_1|$ and $|e_2|$ is if

$$4L - F = 0$$

or

$$L = F/4.$$

The length of the side parallel to the wall is then

$$F - 2L = F/2.$$

This completes the argument (or proof).

The reader familiar with the calculus has seen that what we have done is to go through the steps necessary to evaluate the derivative of A with respect to L. We then set that derivative equal to zero. Requiring that (7) be satisfied for all "very small" values of $|e_1|$ and $|e_2|$ is an intuitive way of "taking a limit."

Actually we have done much more than set the derivative of A equal to zero. If the derivative of a function is zero, we may have a maximum, a minimum, or a point of inflection, depending on the value of the second derivative of the function. In the above argument we actually have

---

4. This is the classic ϵ-δ argument of the calculus. For example, if the student doubts that $4L - F = 0$, ask him to state what value it has. If he says $4L - F = 10^{-10}$, then you choose $e_2 = -10^{-11}$, and inequality (6) is violated. If he chooses a negative value for $4L - F$, say -10^{-100}, then you choose $e_1 = 10^{-101}$, thereby violating (5).

shown that when $L = F/4$, the area, A, is a maximum. (The second derivative is negative.)

It should be pointed out, however, that our argument based on inequalities will only produce the nice result given above if the function to be maximized is a quadratic function (see equation (2)). A more general rigorous solution of maximization problems requires a study of the differential calculus. In this way the study of the calculus can be *motivated* early in the mathematics curriculum.

Other Questions

Notice that if the fencing is in fixed section lengths, S, the above argument breaks down. In particular, although (7) holds, we cannot conclude from (7) that $4L - F = 0$, since we cannot allow $|e_1|$ or $|e_2|$ to be "very small." Indeed neither e_1 nor e_2 can be smaller than the length, S, of one section of fence.

An observant student may ask what happens if we try to make the area as *small* as possible, i.e., to minimize the area. An argument similar to the one above can be given. The result analogous to (7) in this case is

$$(8) \qquad F + |2e_2| < 4L < F - |2e_1|$$

where $e_1 > 0$ and $e_2 < 0$ but the magnitudes of both are arbitrary. In particular, looking at the right-most inequality in (8), as $2e_1$ gets "close" to F, it must be that L gets "close" to zero. However, the left-most inequality in (8) is violated if $L = 0$ since $F + |2e_2|$ clearly is not zero. We seem, therefore, to have reached a contradiction. In fact the second derivative of $A = L(F - 2L)$ is -4, so no minimum exists, and this is the source of our difficulty here.

In our problem it is natural to assume that $L \geq 0$. (We will not allow the length of a side to be negative.) This restriction also makes the problem intractable with the calculus (as did restricting the fencing to fixed section lengths) except by the use of Lagrange multipliers [2].

If we assume that $L \geq 0$, then the right-most inequality in (8) again tells us that $L = 0$. If $L = 0$, the left-most inequality can be ignored since then clearly no negative change, e_2, can be made in L.

The conclusion then is that the area, A, is a minimum and equal to zero when $L = 0$. This is only true, of course, when we restrict L to $L \geq 0$.

One needs to discuss a "mathematical" minimum and a "practical" minimum so that the student understands that, while he can never build an enclosure with zero area, mathematically such an enclosure exists. Moreover, depending on how well he builds fences, he can achieve an area as close to zero as he likes.

A Three-Dimensional Problem

Having analyzed the problem of maximizing an area surrounded by a given length of fencing, we turn to a three-dimensional problem that is analogous to our fencing problem. In terms of programming we will find that instead of a simple loop (see program 1), we will need to write two "nested" loops.

Consider the following problem: We wish to build a rectangular building with a given volume, using as little material as possible for the walls, roof, and floor (see fig. 2).

The volume, V, is given, and

$$(9) \qquad V = S1 \cdot S2 \cdot S3.$$

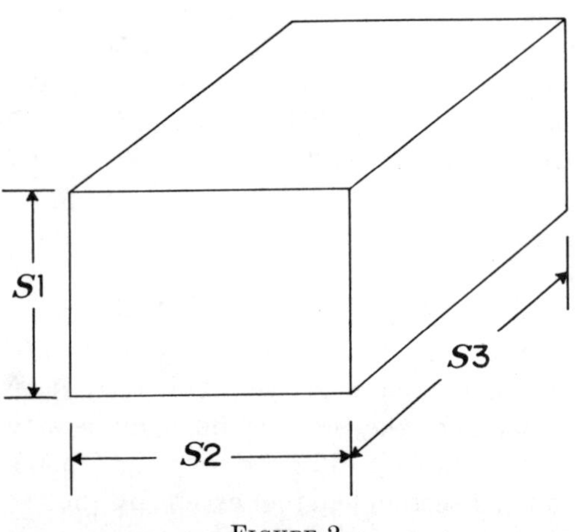

FIGURE 2

The amount of material used for the floor is $S2 \cdot S3$. This is also the amount of material used for the roof. Continuing in this way, the area of the material used to build the structure is

$$(10) \qquad A = 2 \cdot S1 \cdot S2 + 2 \cdot S2 \cdot S3 + 2 \cdot S1 \cdot S3.$$

It is this area that we wish to minimize. Of course, equation (9) must hold.[5]

We will write a program that tests "all" of the various possible dimensions for the building. We note first that there is nothing to distinguish one side from the other. In particular we can interchange any two of $S1$, $S2$, and $S3$ in either or both of equations (9) and (10), and those equations will remain as they are. Because of this symmetry we can treat each side the same as any other. In this connection we note that not all three dimensions can exceed $\sqrt[3]{V}$, for then the right-hand side of (9) would exceed V. Thus at least one of the dimensions must be less than or equal to $\sqrt[3]{V}$. We choose that dimension to be $S1$, the height. We could, of course, have chosen any other side to be the one that does not exceed $\sqrt[3]{V}$ because of the symmetry already noted above; the choice of $S1$ is quite arbitrary.

Our program will let $S1$ be successively $.1, .2, .3, \ldots, \sqrt[3]{V}$; and we will compute the material used for each case. As we shall see, this will not quite be good enough, but it will be a start.

Once we have a value for $S1$, we choose a value for $S2$. Now from (9), for a given value of $S1$,

$$S2 \cdot S3 = V/S1.$$

Again from symmetry considerations we can treat $S2$ and $S3$ alike. Not both of $S2$ and $S3$ can exceed $\sqrt{V/S1}$, for then $S2 \cdot S3$ would exceed $V/S1$. We again arbitrarily choose $S2$ to be less than or

---

5. Notice that we could have started with a given amount of material, i.e. a fixed value for A, and maximized the volume, V. This would have produced the same shape for the building and would have been more closely analogous to the fencing problem discussed earlier in this article. However, we purposely changed to a *minimization* problem to illustrate the "duality" of such optimization problems.

equal to $\sqrt{V/S1}$. Thus we let $S2$ take on the values $.1, .2, \ldots, \sqrt{V/S1}$ for each value of $S1$.

It follows that once we have chosen $S1$ and $S2$ that

$$S3 = V/(S1 \cdot S2),$$

and with these values of $S1$, $S2$, and $S3$ we can compute the material used, A, from equation (10).

The following program (program 2) does just that for the case in which the volume $V = 2$. Notice that in a FOR statement the loop is not executed if the variable being altered exceeds the upper limit of the FOR. Since $\sqrt[3]{V}$ may not be a multiple of .1, we let $U1$, the upper limit of the FOR in statement 30 be $\sqrt[3]{V} + .1$. Similarly we let $U2$, the upper limit on the second FOR statement (statement 50) be $\sqrt{V/S1} + .1$. Notice also that the lower limit on $S2$ is $S1$. There is no need to start $S2$ at .1 since doing so will only repeat a case already calculated. For example, with $V = 2$ when $S1 = .3$, if we let $S2$ start at .1, we would calculate first the case $S1 = .3$, $S2 = .1$, and $S3 = 66.6667$. Then we would calculate $S1 = .3$, $S2 = .2$, $S3 = 33.3333$. However, the values of A corresponding to both of these were calculated before. In particular the area corresponding to the first of these was calculated when $S1 = .1$, $S2 = .3$, and $S3 = 66.6667$. (This was the third calculation made.) The second was calculated when $S1 = .2$, $S2 = .3$, and $S3 = 33.3333$. (This was the second calculation in the second pass through the 30–100 FOR-NEXT loop.)

We could, of course, simply allow $S1$ to vary from .1 to $100 \cdot V$ (the value of $S1$ when $S2 = S3 = .1$), but the computer would then do a lot more work than necessary. Similarly, in statement 50 we could have allowed $S2$ to vary from .1 to $100 \cdot V$ but again the computer would do extra work.

Now, *this program should **not** be run on a computer.* Despite our efforts to keep the amount of work small by allowing $S1$ to

be no larger than $\sqrt[3]{V}$ and $S2$ to be no smaller than $S1$ and no larger than $\sqrt{V/S1}$, this program will print 191 lines of output, each containing four numbers: $S1$, $S2$, $S3$, and A. The computer[6] and terminal[7] time is exorbitant, and we must scan the 191 numbers in the last column ourselves, looking for the smallest number there. It is much better to have the computer determine which of these numbers A is the smallest and then to print the dimensions for that one case.

PROGRAM 2

```
1Ø    READ V
2Ø    LET U1 = V ↑ (1/3) + .1
3Ø    FOR S1 = .1 TO U1 STEP .1
4Ø       LET U2 = SQR(V/S1) + .1
5Ø       FOR S2 = S1 TO U2 STEP .1
6Ø          LET S3 = V/(S1*S2)
7Ø          LET A = 2*(S1*S2+S2*S3+S1*S3)
8Ø          PRINT S1,S2,S3,A
9Ø       NEXT S2
1ØØ   NEXT S1
11Ø   DATA 2
12Ø   END
```

PROGRAM NOTES. *This program should **not** be run on a computer because it produces an excessive amount of output.* $U1$ is the maximum size of the minimum dimension, i.e. not all three dimensions can exceed $U1$. In statement 20 $V \uparrow (1/3)$ is the cube root of V. (The up-arrow indicates exponentiation in BASIC.) The value .1 is added to this because the following FOR will execute the loop down to and including statement 100 as long as $S1$ does not exceed $U1$. If $\sqrt[3]{V}$ is not a multiple of .1, then the loop would not be executed for $S1 = \sqrt[3]{V}$ if we set $U1 = \sqrt[3]{V}$. For a similar reason .1 is added to $\sqrt{V/S1}$ in statement 40. The first FOR-NEXT pair (statements 30 and 100) let $S1 = .1, .2, \ldots, U1$. For *each* of these values of $S1$, the second FOR-NEXT pair (statements 50 and 90) let $S2 = S1, S1 + .1, \ldots, U2$. Given $S1$ and $S2$, statements 60 and 70 compute the area of the material comprising the six sides of the building. Statements 10 and 110 set $V = 2$.

To simultaneously reduce the computer time, the terminal time, and our own bookkeeping work, we modify program 2. We first set M (the minimum area) equal

---

6. The computer time used would be about 13 seconds on the GE Mark I (a 200 series machine) commercial time sharing system.

7. About 20 minutes of terminal time would be used in printing the output on a teletype.

to a large number. Then as we compute each area A, we compare it with M. If A is less than M, we replace M by A. Thus M always contains the value of the smallest area computed to date. If we replace M by A, we also record the dimensions that produced A.

Again we run through the same set of values for $S1$ and $S2$ as we did in program 2. We do not print any results until the entire set of values has been covered. At the end, however, we print M and the dimensions that produced it.

Program 3 does what we have just described in the preceding paragraphs. Notice that the output indicates that the side lengths are *almost* equal.

PROGRAM 3

```
1Ø    READ V
15    LET M = 6*V
2Ø    LET U1 = V ↑ (1/3) + .1
3Ø    FOR S1 = .1 TO U1 STEP .1
4Ø      LET U2 = SQR(V/S1) + .1
5Ø      FOR S2 = S1 TO U2 STEP .1
6Ø        LET S3 = V/(S1*S2)
7Ø        LET A = 2*(S1*S2+S2*S3+S1*S3)
8Ø        IF A > = M THEN 13Ø
9Ø        LET M = A
1ØØ       LET L1 = S1
11Ø       LET L2 = S2
12Ø       LET L3 = S3
13Ø     NEXT S2
14Ø   NEXT S1
15Ø PRINT "SIDE LENGTHS",L1,L2,L3
16Ø PRINT "MATERIAL USED",M
17Ø DATA 2
18Ø END
```

The output from a run of program 3 is as follows:

```
RUN

SIDE LENGTHS    1.2      1.3  1.282Ø5
MATERIAL USED  9.53Ø26
```

PROGRAM NOTES. Starting with program 2 we need type only statement 15 and statements 80 through 180 inclusive.

Statement 10 reads the volume V, which in this case is 2 (see statement 170). Statement 15 sets M equal to a large number. Any number that we are sure is larger than the minimum area will do.

In statement 20 we set the upper limit on $S1$ equal to $\sqrt[3]{V} + .1$. We do this because if $\sqrt[3]{V}$ is not a multiple of .1, we need to try one value larger than $\sqrt[3]{V}$ to make sure we have not skipped one of the possibilities. Similarly, state-

ment 40 sets the upper limit on $S2$ equal to $\sqrt{V/S1} + .1$.

Statement 60 computes $S3$ from equation (9), and statement 70 computes A from equation (10). In statement 80 we compare the area A with the minimum area computed to date, M. If A is not smaller than M, we go to statement 130 and try another value of $S2$. If, however, $A < M$ then we replace M by A (statement 90) and record the dimensions that produced this smaller A in $L1$, $L2$, and $L3$.

This program required about 2.5 seconds of computer time and about 2 minutes of terminal time as compared to the 13 seconds of computer time and 20 minutes of terminal time that program 2 would have used. These reductions are dramatic indeed. Moreover, the computer has actually scanned the column of 191 areas to find the smallest, a task which we would have to do ourselves if we used program 2.

Now the fact that the three dimensions are almost equal leads us to suspect that perhaps if we used steps smaller than .1 we might achieve equal dimensions. Rather than start over with a smaller step size, say .01, we will make use of the results we already have computed.

Since $S1 = 1.2$ when we used steps of .1, it is reasonable to assume that we need only test $S1$ between .1 less and .1 more than 1.2. Thus we will let $S1$ vary from 1.1 to 1.3 in steps of .01. Program 4 does just that.

PROGRAM 4

```
1Ø    READ V
15    LET M = 6*V
3Ø    FOR S1 = 1.1 TO 1.3 STEP .Ø1
4Ø      LET U2 = SQR(V/S1) + .Ø1
5Ø      FOR S2 = S1 TO U2 STEP .Ø1
6Ø        LET S3 = V/(S1*S2)
7Ø        LET A = 2*(S1*S2+S2*S3+S1*S3)
8Ø        IF A > = M THEN 13Ø
9Ø        LET M = A
1ØØ       LET L1 = S1
11Ø       LET L2 = S2
12Ø       LET L3 = S3
13Ø     NEXT S2
14Ø   NEXT S1
15Ø PRINT "SIDE LENGTHS",L1,L2,L3
16Ø PRINT "MATERIAL USED",M
17Ø DATA 2
18Ø END
```

The output from a run of program 4 is:

RUN

SIDE LENGTHS   1.26   1.26 1.25976
MATERIAL USED   9.52441

PROGRAM NOTES: Starting with program 3 only four statements, 20 through 50, need to be typed. Actually statement 20 is deleted.

Notice from the output that to three significant digits, all three dimensions are equal. Moreover, to that accuracy, $S1 = S2 = S3 = \sqrt[3]{V}$.

Notice also that programs 3 and 4 differ in only four statements: 20, 30, 40, and 50. Thus the "conversational" nature of BASIC is vital to our method of solution. After running program 3, we observe that $S1 = 1.2$ and on the basis of that result decide on new bounds, 1.1 and 1.3, for $S1$. Having made this decision, we then type four statements and rerun the program. We could not have written program 4 without first having the results of program 3.[8]

We could now, of course, refine our solution even more by observing that $S1 = 1.26$. Therefore, we will assume $S1$ lies between 1.25 and 1.27. We can rerun the program, letting $S1$ vary from 1.25 to 1.27, in steps of .001 by typing the following.

```
30   FOR S1 = 1.25 TO 1.27 STEP .001
40     LET U2 = SQR(V/S1) + .001
50     FOR S2 = S1 TO U2 STEP .001
```

---

8. The experienced programmer will notice that we could have written our program (program 3) to do this for us. That is, we could have written a third loop surrounding the double loop (statements 15 through 140 inclusive). This third loop initially would set statements 20 to 50 as shown in program 3. The third loop could then reset statements 20 to 50 to those shown in program 4 based on the results of the calculation. This resetting is what we have done "by hand."

There is, of course, always the question of how much of the work should be preplanned and included in the program and how much of the work should be ignored until the need for it arises. The contention here is that in an *educational environment* preplanning should be kept to a minimum. In our problem (minimizing the area for a given volume) we ran into trouble with excessive computing and printing when we did not preplan for computing the minimum area for a given step size, so we did preplan that part of the computation. By failing to preplan the change in step size from .1 to .01 we sacrifice very little but simplify the programming work by a considerable amount. Therefore, we choose not to preplan and not to program the step size change.

And then

RUN

This refinement process could be carried further if we desire to do so.

The important point is that *we use the results of one run to determine the program for the next run*. Without a conversational capability this process is quite cumbersome. With a conversational capability it is extremely simple and straightforward.

Integral Dimensions

For arbitrary dimensions, it turns out that in fact all three dimensions are identical, equalling $\sqrt[3]{V}$. We shall not prove this. An argument similar to that given earlier, under "Verification of a Conjecture," can be given, but it is cumbersome. A proof is best left until after a study of the calculus when it becomes trivial.

If we insist that all dimensions be integers, however, the problem becomes more difficult to manage and is intractable through the calculus. Our program, however, can handle the problem with only minor changes.

To do so, we let $S1 = 1, 2, \ldots, \sqrt[3]{V}$. If $\sqrt[3]{V}$ is not an integer, then we can stop at the largest integer less than $\sqrt[3]{V}$ since $S1$ cannot exceed $\sqrt[3]{V}$. For each value of $S1$ we let $S2 = S1, S1 + 1, \ldots, \sqrt{V/S1}$.

Now again we compute $S3$ from equation (9). However, $S3$ may not be an integer even though V, $S1$, and $S2$ are integers. If $S3$ is not an integer, then we need not calculate A.

Program 5 carries out the calculations. It uses the INT function which computes the integer part of a variable. If x is an integer, then $INT(x) = x$. Otherwise, $INT(x) \neq x$. Statement 65 in program 5 says that if $S3 \neq INT(S3)$ then we should skip to statement 130 and try another value of $S2$.

PROGRAM 5

```
10   READ V
15   LET M = 6*V
20   LET U1 = V ↑ (1/3)
30   FOR S1 = 1 TO U1
40     LET U2 = SQR(V/S1)
```

```
50      FOR S2 = S1 TO U2
60        LET S3 = V/(S1*S2)
65        IF S3 < > INT(S3) THEN 130
70        LET A = 2*(S1*S2+S2*S3+S1*S3)
80        IF A > = M THEN 130
90        LET M = A
100       LET L1 = S1
110       LET L2 = S2
120       LET L3 = S3
130     NEXT S2
140   NEXT S1
150   PRINT "SIDE LENGTHS",L1,L2,L3
160   PRINT "MATERIAL USED",M
170   DATA 30
180   END
```

The output from a run of program 5 is:

```
RUN

SIDE LENGTHS      2      3      5
MATERIAL USED     62
```

PROGRAM NOTES: Starting from either program 3 or program 4 we need type only six statements, 20, 30, 40, 50, 65, and 170. In statement 65 the symbol $<>$ means "less than *or* greater than" which is equivalent to "not equal."

Notice that for $V = 30$ the dimensions are $S1 = 2$, $S2 = 3$, and $S3 = 5$. Clearly these could not be obtained by using $\sqrt[3]{V}$ and rounding off to the nearest integer since $\sqrt[3]{30} = 3.107233$.

Another Three-Dimensional Problem

We close our discourse with one more variation of the building problem.

Suppose we are building a storage shed of a given volume, and we are willing to use the ground as the floor. Then we need build only *five* rectangular areas (four sides and a roof). What is the shape that will minimize the amount of material needed to build the shed?

We return to program 3 (the first reasonable program for the original six-sided building problem). It is quite easy to change program 3 so that it solves this new five-sided problem. In particular we need change only three statements: 20, 30, and 70. Actually we must also change statement 170 to use different data, and we change the PRINT statements.

Referring back to figure 2 we see that now all three sides cannot be treated in the same way since there is no bottom (or

floor). In other words, although equation (9) for the volume is symmetric in $S1$, $S2$, and $S3$, the equation for the area A will not be symmetric. Indeed, equation (10) is replaced by

$$(11) \quad A = 2 \cdot S1 \cdot S2 + S2 \cdot S3 + 2 \cdot S1 \cdot S3.$$

This differs from equation (10) in the second term on the right-hand side. While we can interchange $S2$ and $S3$ in both equations (9) and (11) without disturbing anything, we cannot interchange either of these with $S1$ without altering equation (11) itself.

Thus we cannot say that $S1$ will not exceed $\sqrt[3]{V}$ as we did previously. We will somewhat arbitrarily assume that $S2$ and $S3$ will not be smaller than 1, and thus that $S1$ will not exceed V.[9] Thus we change statement 30 in program 3 to read as follows:

```
30 FOR S1 = .1 TO V STEP .1
```

Since equations (9) and (11) are symmetric in $S2$ and $S3$, we need not change statement 40. Statement 70 is changed to reflect equation (11). The resulting program, program 6, is shown below together with its output.

```
                  PROGRAM 6
10      READ V
15      LET M = 6*V
30      FOR S1 = .1 TO V STEP .1
40        LET U2 = SQR(V/S1) + .1
50        FOR S2 = .1 TO U2 STEP .1
60          LET S3 = V/(S1*S2)
70          LET A = 2*S1*(S2+S3)+S2*S3
80          IF A > = M THEN 130
90          LET M = A
100         LET L1 = S1
110         LET L2 = S2
120         LET L3 = S3
130       NEXT S2
140     NEXT S1
150     PRINT "HEIGHT IS",L1
155     PRINT "SIDE LENGTHS",L2,L3
160     PRINT "MATERIAL USED",M
170     DATA 4
180     END
```

```
RUN

HEIGHT IS          1.
```

---

9. If we allow both $S2$ and $S3$ to be as small as .1, then statement 30 should read: 30 FOR $S1 = .1$ TO $100*'V$ STEP .1.

```
SIDE LENGTHS      2.        2.
MATERIAL USED    12.
```

PROGRAM NOTES: Starting with program 3 only six statements need to be typed. They are statements 20, 30, 70, 150, 155, and 170. Statement 20 is deleted. Statement 170 merely changes the data (volume, V) from 2 to 4. Statements 150 and 155 arrange that the height, $S1$, be distinguished from the other two dimensions.

Notice that in this case the two lateral dimensions, $S2$ and $S3$, are *twice* the height, $S1$. Recall that in the fencing problem described previously the length of the side parallel to the wall was *twice* the length of the side perpendicular to the wall.

Before concluding that $S2$ and $S3$ will always be twice $S1$, let us try another volume, say $V=2$. To do so we merely type the following, and then rerun the program.

170 DATA 2

The result is shown in program 7.

PROGRAM 7

```
10   READ V
15   LET M = 6*V
30   FOR S1 = .1 TO V STEP .1
40     LET U2 = SQR(V/S1) + .1
50     FOR S2 = .1 TO U2 STEP .1
60       LET S3 = V/(S1*S2)
70       LET A = 2*S1*(S2+S3)+S2*S3
80       IF A > = M THEN 130
90       LET M = A
100      LET L1 = S1
110      LET L2 = S2
120      LET L3 = S3
130    NEXT S2
140  NEXT S1
150  PRINT "HEIGHT IS",L1
155  PRINT "SIDE LENGTHS",L2,L3
160  PRINT "MATERIAL USED",M
170  DATA 2
180  END
```

The output from a run of program 7 is as follows:

```
RUN

HEIGHT IS          .8
SIDE LENGTHS     1.5        1.66667
MATERIAL USED    7.56667
```

PROGRAM NOTES: This differs from program 6 only in the DATA statement, 170.

In program 7, $S2$ and $S3$ are almost twice $S1$. It is a little disturbing that $S2 \neq S3$, since our symmetry argument would lead us to believe that $S2$ and $S3$ should be the same. To resolve these problems, we follow the pattern that we followed in a previous case—that is, we try a smaller step size, say .01.

Again we note that the results of program 7 indicate that $S1$ is approximately .8. Therefore, we let $S1$ vary from .7 to .9 in steps of .01. To do this we rewrite statement 30.

30 FOR S1 = .7 TO .9 STEP .01

Similarly, since program 7 indicates that $S2$ is approximately 1.5, we type the following.

50 FOR S2 = 1.4 TO 1.6 STEP .01

We then rerun the program. (Of course, we could delete statement 40 since it no longer contributes to the program, but it does no harm to allow it to stay.) This new program together with its results appears as program 8.

PROGRAM 8

```
10   READ V
15   LET M = 6*V
30   FOR S1 = .7 TO .9 STEP .01
40     LET U2 = SQR(V/S1) + .1
50     FOR S2 = 1.4 TO 1.6 STEP .01
60       LET S3 = V/(S1*S2)
70       LET A = 2*S1*(S2+S3)+S2*S3
80       IF A > = M THEN 130
90       LET M = A
100      LET L1 = S1
110      LET L2 = S2
120      LET L3 = S3
130    NEXT S2
140  NEXT S1
150  PRINT "HEIGHT IS",L1
155  PRINT "SIDE LENGTHS",L2,L3
160  PRINT "MATERIAL USED",M
170  DATA 2
180  END
```

```
RUN

HEIGHT IS          .79
SIDE LENGTHS     1.59       1.59223
MATERIAL USED    7.55957
```

PROGRAM NOTES: Starting with program 7 we need only type statements 30 and 50 to arrive at program 8. We can also delete statement 40 from program 8 if we wish to do so.

Now rounded off to the nearest tenth,

the results of program 8 indicate that $S2 = S3 = 2 \cdot S1 = \sqrt[3]{4}$. We can carry the process one step further by letting $S1$ vary from .78 to .79 in steps of .001 and also letting $S2$ vary from 1.58 to 1.60 in steps of .001. This is done in the final program, program 9. The results show that, to the nearest hundredth, $S2 = S3 = 2 \cdot S1 = \sqrt[3]{4}$.

PROGRAM 9

```
1Ø    READ V
15    LET M = 6*V
3Ø    FOR S1 = .78 TO .8Ø STEP .001
4Ø      LET U2 = SQR(V/S1) + .1
5Ø      FOR S2 = 1.58 TO 1.6Ø STEP .ØØ1
6Ø        LET S3 = V/(S1*S2)
7Ø        LET A = 2*S1*(S2+S3)+S2*S3
8Ø        IF A > = M THEN 13Ø
9Ø        LET M = A
1ØØ       LET L1 = S1
11Ø       LET L2 = S2
12Ø       LET L3 = S3
13Ø     NEXT S2
14Ø   NEXT S1
15Ø PRINT "HEIGHT IS",L1
155 PRINT "SIDE LENGTHS",L2,L3
16Ø PRINT "MATERIAL USED",M
17Ø DATA 2
18Ø END
```

The output from a run of program 9 is as follows:

```
RUN

HEIGHT IS          .794
SIDE LENGTHS       1.587      1.5872
MATERIAL USED      7.55953
```

PROGRAM NOTES: Starting with program 8 only statements 30 and 50 need to be typed.

Notice that the statements in program 9 depended on the results of program 8. Program 8's statements similarly depended on the results of program 7. Thus we relied heavily on the "conversational" nature of BASIC. This way of utilizing BASIC's conversational capability was, of course, encountered earlier, when the statements in program 4 depended on the results of program 3. As we pointed out in footnote 8, with enough preplanning and

some more programming we could have the computer generate the sequence of programs 7, 8, and 9. However, in this case the effort involved in such preplanning seems to be more than the results would warrant.

Returning to the problem stated at the beginning of this section, we now have the results for two values of the volume, V. If $V = 4$, $S2 = S3 = 2 \cdot S1 = \sqrt[3]{8}$. If $V = 2$, $S2 = S3 = 2 \cdot S1 = \sqrt[3]{4}$. This leads us to *conjecture* that the following equation holds for *any* V.

$$(12) \qquad S2 = S3 = 2 \cdot S1 = \sqrt[3]{2 \cdot V}$$

The student may wish to try other values of V to verify this conjecture.

The conjecture in equation (12) is indeed a correct one. It can be verified by the methods described earlier in this article. However, the algebra and logic become rather tedious. For that reason the verification of this conjecture is best left until after the calculus has been studied. With the calculus available, of course, the verification becomes trivial.

We should, therefore, have *motivated* the student to study the calculus so that he may verify not only this conjecture but other similar ones that arise in maximization and minimization problems.

BIBLIOGRAPHY

1. Committee on Computer-oriented Mathematics. *Introduction to an Algorithmic Language (BASIC)*. Washington, D.C.: National Council of Teachers of Mathematics, 1963.
2. DORN, WILLIAM S. "On Lagrange Multipliers and Inequalities." *Operations Research* 9 (January–February 1961): 95–104.
3. DORN, WILLIAM S., and JUDITH B. EDWARDS. "Finding the 'Best' Solution via Computer." *Journal of Educational Data Processing* 6 (Spring 1969): 90–107.
4. KEMENY, JOHN G., and THOMAS E. KURTZ. *BASIC Programming*. New York: John Wiley & Sons, 1967.

Use of DIRECTIONAL DERIVATIVE in Locating Extrema

By **WILLIAM C. STRETTON**

College of DuPage

Glen Ellyn, Illinois

IT IS possible to use the directional derivative effectively in locating relative extrema of a function of two variables. Why this method has been virtually overlooked is a mystery. The criteria given in most texts, if given at all, for testing whether a critical point is a maximum or a minimum are usually developed from a Taylor's series for functions of two variables. It has been my experience that most first-year calculus students get lost somewhere in this development. The method I am going to propose employs methods used in functions of a single variable and hence is more meaningful, I've found, with such students.

There are two approaches one can take. The first is to develop an expression for the second directional derivative and use this expression in each individual problem. The second is to use this expression for the second directional derivative to develop the previously mentioned criteria. We shall look at both approaches in the order stated.

The directional derivative for a function of two variables, $w = f(x, y)$, at the point (a, b) in a direction θ with the positive x-axis, is defined as

$$(1) \qquad \frac{dw}{ds} = \lim_{\Delta s \to 0} \frac{\Delta w}{\Delta s}.$$

Here s is measured along the line ℓ in the xy-plane in the given direction (see fig. 1). Since x and y can be represented parametrically in terms of s (see fig. 2), we can use the chain rule to derive from the definition the common expression for the directional derivative, namely,

$$(2) \qquad \frac{dw}{ds} = f_x(a, b) \cos \theta + f_y(a, b) \sin \theta.$$

Now geometrically,

$$\frac{dw}{ds}$$

represents the slope of a curve that is the

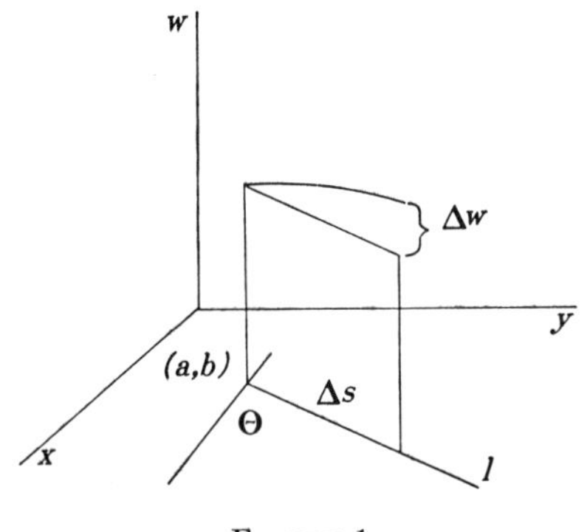

FIGURE 1

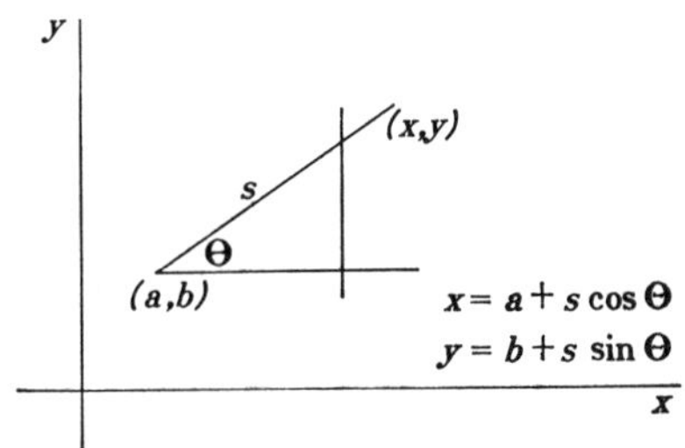

FIGURE 2

139

trace of some plane through ℓ perpendicular to the xy-plane in the surface $w = f(x, y)$. Furthermore,

$$\frac{d^2w}{ds^2}$$

will tell us whether this curve is concave upward or downward, depending on whether

$$\frac{d^2w}{ds^2} > 0$$

or

$$\frac{d^2w}{ds^2} < 0.$$

All we need then is an expression for

$$\frac{d^2w}{ds^2}.$$

This is readily obtained through use of the chain rule again. Let

$$w' = \frac{dw}{ds}.$$

Then, since w' can be considered a function of s alone, we have

$$\frac{dw'}{ds} = \frac{\partial w'}{\partial x} \frac{dx}{ds} + \frac{\partial w'}{\partial y} \frac{dy}{ds}.$$

But

$$\frac{\partial w'}{\partial x} = \frac{\partial^2 f}{\partial x^2} \cos \theta + \frac{\partial^2 f}{\partial x \partial y} \sin \theta,$$

$$\frac{\partial w'}{\partial y} = \frac{\partial^2 f}{\partial y \partial x} \cos \theta + \frac{\partial^2 f}{\partial y^2} \sin \theta,$$

and

$$\frac{dx}{ds} = \cos \theta,$$

$$\frac{dy}{ds} = \sin \theta.$$

Therefore, we have

$$(3) \quad \frac{dw'}{ds} = \frac{d^2w}{ds^2}$$

$$= \left(\frac{\partial^2 f}{\partial x^2} \cos \theta + \frac{\partial^2 f}{\partial x \partial y} \sin \theta \right) \cos \theta$$

$$+ \left(\frac{\partial^2 f}{\partial y \partial x} \cos \theta + \frac{\partial^2 f}{\partial y^2} \sin \theta \right) \sin \theta$$

$$= \frac{\partial^2 f}{\partial x^2} \cos^2 \theta + 2 \frac{\partial^2 f}{\partial x \partial y} \sin \theta \cos \theta$$

$$+ \frac{\partial^2 f}{\partial y^2} \sin^2 \theta,$$

assuming that

$$\frac{\partial^2 f}{\partial x \partial y} = \frac{\partial^2 f}{\partial y \partial x}.$$

Now suppose we have found a point (a, b) at which

$$f_x(a, b) = f_y(a, b) = 0.$$

Then if

$$\left. \frac{d^2w}{ds^2} \right|_{(a,b)} = f_{xx}(a, b) \cos^2 \theta$$

$$+ 2 f_{xy}(a, b) \sin \theta \cos \theta$$

$$+ f_{yy}(a, b) \sin^2 \theta$$

is greater than zero for all θ, we know that all traces by planes through (a, b) perpendicular to the xy-plane are concave upward and so we must have a relative minimum at (a, b). Similarly, if

$$\frac{d^2w}{ds^2} < 0$$

at (a, b) for all θ, we have a relative maximum. If

$$\frac{d^2w}{ds^2}$$

is positive for some values of θ and negative for others at (a, b), then we know we have a saddle point at (a, b). However, the test will fail if either

$$\frac{d^2w}{ds^2} = 0$$

for all θ or if

$$\frac{d^2w}{ds^2} = 0$$

for some θ and is positive (negative) for all other values of θ.

In practice it is usually quite easy to tell how the sign of

$$\frac{d^2w}{ds^2}$$

behaves at a given point. Let us illustrate with an example. Consider the function

$$w = x^3 + y^3 - 6xy + 3.$$

We get $f_x = 3x^2 - 6y$ and $f_y = 3y^2 - 6x$. Setting these equal to zero and solving simultaneously we find the critical points are $(0, 0)$ and $(2, 2)$. Now $f_{xx} = 6x$, $f_{yy} = 6y$, and $f_{xy} = -6$. At $(0, 0)$ we have

$$\frac{d^2w}{ds^2} = -12 \sin \theta \cos \theta = -6 \sin 2\theta.$$

As $\sin 2\theta$ is sometimes positive and sometimes negative, so

$$\frac{d^2w}{ds^2}$$

will be sometimes negative and sometimes positive. Hence there is a saddle point at $(0, 0)$. At $(2, 2)$,

$$\frac{d^2w}{ds^2} = 12 \cos^2 \theta - 6 \sin 2\theta + 12 \sin^2 \theta$$

$$= 12 - 6 \sin 2\theta.$$

Since the maximum value of $\sin 2\theta$ is 1, it follows that

$$\frac{d^2w}{ds^2} > 0$$

for all θ at $(2, 2)$ and so there is a relative minimum at $(2, 2)$.

The second approach, mentioned at the beginning, is to use the expression (3) to develop certain well known criteria. To this end it is convenient to write (3) in two alternate forms. By completing squares and other algebraic manipulations (3) can be written as

$$(4) \qquad \frac{d^2w}{ds^2} = f_{xx} \left[\left(\cos \theta + \frac{f_{xy} \sin \theta}{f_{xx}} \right)^2 \right.$$

$$\left. + \frac{\sin^2 \theta}{f_{xx}^2} (f_{xx}f_{yy} - f_{xy}^2) \right]$$

or

$$(5) \qquad \frac{d^2w}{ds^2} = f_{yy} \left[\left(\sin \theta + \frac{f_{xy} \cos \theta}{f_{yy}} \right)^2 \right.$$

$$\left. + \frac{\cos^2 \theta}{f_{yy}^2} (f_{xx}f_{yy} - f_{xy}^2) \right]$$

Now let $\Delta = f_{xx}f_{yy} - f_{xy}^2$. Then it can be seen from (4) or (5) that $\frac{d^2w}{ds^2}$ will agree in sign with f_{xx} or f_{yy} if $\Delta > 0$. (If $\Delta > 0$, f_{xx} and f_{yy} must have the same sign.) So, at a point (a, b) where $f_x(a, b) = f_y(a, b) = 0$, there will be a relative maximum if $\Delta > 0$ and $f_{xx} < 0$, and a relative minimum if $\Delta > 0$ $f_{xx} > 0$.

If $\Delta < 0$ and $f_{xx} \neq 0$ at (a, b), (4) shows that

$$\frac{d^2w}{ds^2}$$

agrees in sign with f_{xx} whenever $\sin \theta = 0$ but has the opposite sign when

$$\cos \theta + \frac{f_{xy} \sin \theta}{f_{xx}} = 0,$$

that is, when

$$\tan \theta = -\frac{f_{xx}}{f_{xy}}.$$

If $f_{xx} = 0$ but $f_{yy} \neq 0$, the same reasoning can be used on (5) to show that

$$\frac{d^2w}{ds^2}$$

changes sign. If both f_{xx} and f_{yy} are zero, then (3) shows that

$$\frac{d^2w}{ds^2} = f_{xy} \sin 2\theta$$

so that

$$\frac{d^2w}{ds^2}$$

changes sign as $\sin 2\theta$ changes sign. Hence in all cases where $\Delta < 0$ at a critical point,

$$\frac{d^2w}{ds^2}$$

will alternate in sign as θ varies so there will be neither a relative maximum nor a relative minimum but a saddle point.

Finally, if $\Delta = 0$, it can be shown that either

$$\frac{d^2w}{ds^2} = 0$$

for all θ, or else

$$\frac{d^2w}{ds^2}$$

agrees in sign with f_{xx} (or f_{yy}) for some values of θ but is zero for some values of θ. (I leave it to the reader to supply the details.) In either case no definite conclusion as to maximum, minimum, or saddle point can be drawn if $\Delta = 0$ at a critical point.

To summarize, in trying to locate relative extrema of a function of two variables, $w = f(x, y)$, one can use the expression developed for

$$\frac{d^2w}{ds^2}$$

at each critical point, that is, each point where $f_x(x, y) = f_y(x, y) = 0$. Then if

$$\frac{d^2w}{ds^2} > 0$$

for all θ at that point, there is a relative minimum; if

$$\frac{d^2w}{ds^2} < 0$$

for all θ, there is a relative maximum. Or, one can use the expression for

$$\frac{d^2w}{ds^2}$$

to develop the following criteria for application at any point (a, b) where $f_x(a, b) = f_y(a, b) = 0$:

Let $\Delta = f_{xx}f_{yy} - f_{xy}^2$. Then
1. If $\Delta > 0$ and $f_{xx}(a, b) < 0$, $f(a, b)$ is a relative maximum.
2. If $\Delta > 0$ and $f_{xx}(a, b) > 0$, $f(a, b)$ is a relative minimum.
3. If $\Delta < 0$, $f(a, b)$ is a saddle point.
4. If $\Delta = 0$, no conclusion can be drawn.

Needless to say, in any search for extrema one must also examine those points, if any, at which the partials fail to exist. Obviously the above techniques apply only at points where there are well-defined partial derivatives.

Bibliography: *Maxima and Minima*

Hotelling, Harold. Some Little Known Applications of Mathematics. 29 (April 1936): 157–69.

Expository article covering the physical, biological, and social sciences, as well as literature. Includes remarks on the application of calculus to various maxima and minima problems in economics.

Tierney, John A. Elementary Techniques in Maxima and Minima. 46 (November 1953): 484–86.

Remarks on some maxima and minima problems solved without derivatives.

Lange, L. H. Some Inequality Problems. 56 (November 1963): 490–94.

Remarks on maxima and minima problems solved without derivatives.

Nannini, Amos. Maxima and Minima by Elementary Methods. 60 (January 1967): 31–32.

Remarks on maxima and minima problems solved without derivatives.

Kovach, Ladis D. A Note on Curve-Fitting with Rational Polynomials. 60 (February 1967): 129–32.

Remarks concerning the problem of fitting a curve to given data. Includes comments on the method of least squares.

Viertel, William K. A Visual Aid for an Elementary Applied Maximum Problem in Calculus. 61 (January 1968): 29–30.

Step-by-step directions for the construction of a visual aid useful in illustrating an applied maximum problem.

Bird, M. T. Maximum Rectangle Inscribed in a Triangle. 64 (December 1971): 759–60.

Some additional remarks on the article by Lange (listed above).

Fabricant, Mona. A Classroom Discovery in High School Calculus. 65 (December 1972): 744–45.

Remarks on a student experience-discovery approach to teaching maxima and minima.

6

Indeterminate Forms

Many calculus students find it extremely difficult to accept such concepts as "the limit of a function of the form $0/0$ can be some nonzero constant" or "the limit of a function of the form 1^∞ can be some number other than one." The articles in this section suggest some possible means to clarify these ideas.

Baylock, in simple language, presents a graphical approach to visualizing the limit of an indeterminate function. Next, Williams discusses the use of a computer to demonstrate numerically the limiting process for indeterminate forms.

Through such concrete illustrations, the student may be led to a greater appreciation and understanding of the abstract calculus concepts involved. Thus, some of the mystery frequently associated with indeterminate forms may be dispelled.

Graphical interpretation of the limit of an indeterminate function

by Adrian Baylock, Saint Francis College, Loretto, Pennsylvania

One of the first concepts encountered in the calculus, and surely the most fundamental, is that of a limit. Very early in the process of evaluating limits of various functions, the student meets such expressions as

$$\lim_{x \to 2} \frac{x^2-4}{x-2}$$

which, by applying the methods he has been taught up to that time for finding limits, yields the result $\frac{0}{0}$ at $x=2$. We then explain that this result is "indeterminate," and since it is not advisable to introduce L'Hospital at this early stage, we show certain techniques that enable the student to find a limit for this type of function. The usual procedure is to factor the numerator. Thus,

$$\lim_{x \to 2} \frac{x^2-4}{x-2} = \lim_{x \to 2} \frac{(x+2)(x-2)}{x-2} .$$

Assume $x \neq 2$ and reduce the expression by $(x-2)$

$$\lim_{x \to 2} \frac{x^2-4}{x-2} = \lim_{x \to 2} \frac{(x+2)(x-2)}{x-2} ;$$

that is,

$$\frac{x^2-4}{x-2} = x+2$$

for all values of x except $x=2$.

When explained, it is usually vaguely clear to the student what is meant by say-ing that the function in its original form is indeterminate at $x=2$, but not so clear why its limit should be 4. At this point in the student's development, one picture is worth ten thousand words. A graph of the function shows clearly why the limit is 4.

To illustrate, consider the graph of the function cited above. Let

$$y = f(x) = \frac{x^2-4}{x-2} .$$

This expression is equal to a unique real number for all values of x except $x=2$, so in the following discussion we will make the assumption that $x \neq 2$.

Then we may write:

$$y(x-2) = (x+2)(x-2) \tag{1}$$
$$y(x-2) - (x+2)(x-2) = 0 \tag{2}$$

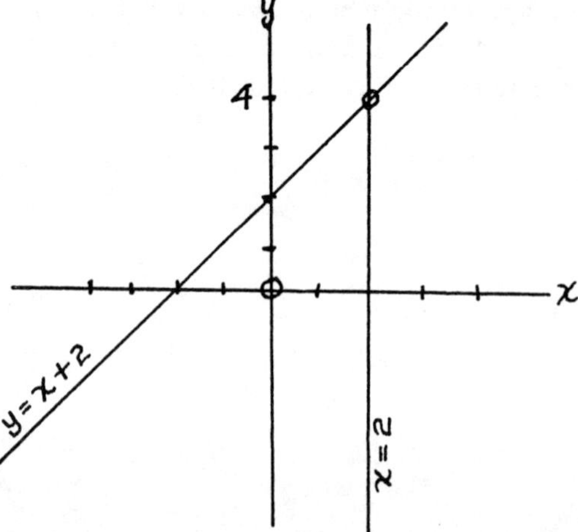

Figure 1

$$[x-2][y-(x+2)]=0 \qquad (3)$$

$$x-2=0 \text{ or } x=2 \qquad (4)$$

$$y-x-2=0 \text{ or } y=x+2 \qquad (5)$$

Graphing the equations (4) and (5), we have the two straight lines shown in Figure 1.

Bearing in mind that the two lines together make up the graph, it is clear that the expression

$$\frac{x^2-4}{x-2}$$

has a value for every value of x. But at $x=2$, it has too many values. When $x=2$, the expression could have any value from $-\infty$ to $+\infty$, since the line $x=2$ is, of course, infinite. We say the expression is "indeterminate" at $x=2$. It is evident from the graph that the function approaches the limit 4 as x approaches 2 for values of x both greater than 2 and less than 2.

This graphical "proof" that

$$\lim_{x\to 2} \frac{x^2-4}{x-2}$$

should be 4 appeals to the student's intuition and satisfies him of the "naturalness" of expressions such as $\frac{0}{0}$ at the same time.

A demonstration of indeterminate forms using finite methods

HORACE E. WILLIAMS, *Vanderbilt University, Nashville, Tennessee.*
The teaching of traditional units in this area of calculus
may be facilitated by the use of numerical demonstrations.

WHEN STUDYING limits of functions in traditional courses in elementary calculus, one generally considers a unit on "Indeterminate Forms." In these units functions which approach indeterminate limits, symbolized as "$\frac{0}{0}$, $\frac{\infty}{\infty}$, $\infty - \infty$, $0 \cdot \infty$, 0^0, 1^∞," are characteristically encountered.* To evaluate these indeterminate forms, the theorems known as "L'Hospital's Rules" are developed and applied with modifications to the several situations.

As in any situation involving the limit concept, students may be skeptical that the functions actually approach the limits predicted by the L'Hospital's-Rule solutions. This may be due to a hazy understanding of what is actually taking place in the all-important process of evaluating limits of functions. Something more than reviewing the definition of limit is often needed.

As examples of this, many teachers are annually confronted with questions of the following types from their calculus students: "Do not functions of the form $\frac{0}{0}$ always approach 0?"; or, "Are not functions of the form $\frac{0}{0}$ always undefined?"; or, "How can a function of the form 1^∞ approach any limit other than 1, since 1 to any power is 1 by definition?"

With the firm conviction in mind that a thorough and concrete understanding of elementary limits constitutes an invaluable asset to complete mastery and effective application of college calculus, it was decided to develop, as concrete teaching aids, numerical demonstrations of the limiting process in action for several selected "Indeterminate Form" exercises, using an IBM 7072 computer to perform the desired calculations. The exercises, common to many elementary calculus texts, that were selected for the numerical demonstrations are as follows:

$$\lim_{x \to 0} \frac{e^x - e^{-x}}{\tan x},$$

$$\left(\text{type } \frac{0}{0}, \text{ value of limit is 2}\right) \quad (1)$$

$$\lim_{x \to 0} \frac{1}{\sin^2 x} - \frac{1}{x^2},$$

$$(\text{type } \infty - \infty, \text{ value of limit is } \tfrac{1}{3}) \quad (2)$$

$$\lim_{x \to 0} (1 + \tan x)^{\frac{1}{x}},$$

$$(\text{type } 1^\infty, \text{ value of limit is } e) \quad (3)$$

$$\lim_{x \to 0} (\sin x)^{\tan x},$$

$$(\text{type } 0^0, \text{ value of limit is 1}) \quad (4)$$

Each of the functions given in (1), (2), (3), and (4) was evaluated first for $x = 1.00$ and then evaluated with successive increments of $-.01$ down to $x = .01$. Next, each function was evaluated starting at $x = .01$ for successive arguments of $-.0001$ down to $x = .0001$. The values for the arguments and the functions were printed for each recursion so that the students might see the numerical behavior of these as the

---

* George B. Thomas, Jr., *Calculus and Analytic Geometry* (Reading, Mass.: Addison-Wesley Publishing Co., Inc., 1960), pp. 814–821.

argument approached 0 from a starting value of $x = 1.0$.

Since the fractions in (1) and (2) may be shown to be even functions, allowing x to approach 0 using negative arguments would reproduce the same values of the fractions for cases where the absolute values of the arguments are equal. Therefore, (1) and (2) were not computed for negative arguments. The function (4) was not evaluated for negative arguments, since this involved going beyond real variable functions and was considered outside the scope of the demonstration. Function (3) was, however, evaluated, starting at $x = -0.50$ and going to $x = -0.01$ using increments of $+.01$, and from -0.01 down to -0.0001 using increments of $+.0001$.

Table 1 gives selected values for these functions and the arguments used in their computations.

Several observable facets of the demonstrations seemed appealing to students. It was interesting to notice how some functions [notably (2)] converged quite rapidly on the limit value, while others [notably (3) and (4)] converged rather slowly and at first were actually moving away from the limit value. It was also interesting from the numerical analysis standpoint to observe the functions converging to a point of maximum accuracy. Having the students construct graphs of these functions, on a highly blown-up scale, as they approach their respective limiting values is a quite useful exercise for dramatizing the aforementioned characteristics.

TABLE 1

x	$\dfrac{e^x - e^{-x}}{\tan x}$	$\dfrac{1}{\sin^2 x} - \dfrac{1}{x^2}$	$(1 + \tan x)^{\frac{1}{x}}$	$(\sin x)^{\tan x}$	$(1 + \tan y)^{\frac{1}{y}}$ where $y = -x$
1.0000	1.509176	0.412283	2.557408	0.764285	
0.9000	1.629187	0.395156	2.474501	0.735109	
0.8000	1.725083	0.380758	2.422553	0.710328	
0.7000	1.801245	0.368727	2.393769	0.690479	
0.6000	1.861188	0.358777	2.383941	0.676362	
0.5000	1.907717	0.350685	2.391051	0.669233	4.8581084
0.4000	1.943041	0.344277	2.414651	0.671167	3.9506762
0.3000	1.968863	0.339420	2.455673	0.685857	3.4338220
0.2000	1.986444	0.336017	2.516544	0.720649	3.1039770
0.1000	1.996653	0.334000	2.601642	0.793584	2.8786588
0.0900	1.997291	0.333870	2.611732	0.804589	2.8601490
0.0700	1.998364	0.333660	2.632886	0.829850	2.8249319
0.0500	1.999166	0.333520	2.655405	0.860766	2.7919613
0.0300	1.999698	0.333400	2.679376	0.900115	2.7610672
0.0100	1.999973	0.333000	2.704996	0.954991	2.7320962
0.0090	1.9999682	0.3330000	2.7062006	0.9584901	**
0.0080	1.9999823	0.3330000	2.7075096	0.9621091	**
0.0070	1.9999931	0.3340000*	2.7087153	0.9658628	**
0.0060	1.9999777	*	2.7101603	0.9697700	**
0.0050	1.9999853	*	2.7115024	0.9738560	2.7251473
0.0040	1.9999918	*	2.7128451	0.9781561	2.7237672
0.0030	1.9999974	*	2.7141884	0.9827235	2.7224086
0.0020	2.0000024*	*	2.7155325	0.9876477	2.7210404
0.0010	*	*	2.7168146	0.9931160	2.7197563
0.0008	*	*	2.7170648	0.9943115	2.7194901
0.0006	*	*	2.7172732	0.9955587	2.7192552
0.0004	*	*	2.7175341	0.9968753	2.7189423*
0.0002	*	*	2.7176905	0.9982980	2.7189423*
0.0001	*	*	2.7170648*	0.9990794	2.718942*

* Point of maximum accuracy reached. Thereafter, machine error precluded further convergence of the functions.

** Values not computed.

Bibliography: *Indeterminate Forms*

Tripp, M. O. Indeterminate Forms in Trigonometry. 11 (March 1919): 118–20.
Remarks on possible pitfalls in the use of trigonometric operations.

Struik, D. J. The Origin of L'Hôpital's Rule. 56 (April 1963): 257–60. Some historical remarks concerning L'Hôpital's rule.

Ash, Carol. A Plausibility Argument for L'Hôpital's Rule. 61 (April 1968): 403–4.
Remarks on an approach to teaching L'Hôpital's rule.

Kingston, J. Maurice. More about 0^0. 64 (May 1971): 447.
Counterexamples to a proposed definition that $0^0 = 1$.

Tebbs, Richard Ray. More about 0^0. 64 (December 1974): 742.
Remarks on a means of constructing a class of indeterminate functions that will approach a specified limit.

7

Integration

Integration theory and applications present many pedagogical challenges for the calculus teacher. The material is often more difficult for beginning students to grasp than the theory of differentiation. The articles appearing in this section offer some classroom-tested suggestions for teaching this part of the calculus curriculum.

Part one of this section deals with finding plane areas by approaches not involving integration. A theorem from solid geometry is the basis for the method proposed by Pursell. Morgan gives a summation approach.

In part two, integration techniques and special integrals are dealt with. Hellman presents an approach for evaluating $\int \sec x \, dx$ and $\int \csc x \, dx$. Rogers and Rogers discuss the partial-fraction approach to integration.

Several integration applications are considered in the third part. Calculation of the moment of inertia is examined by Lehpamer. The article by Atneosen deals with the area of curved surfaces.

This section illustrates some of the uses of integration. It also points up the need to develop methods for evaluating the integrals that may arise. The material presented here has either direct applicability to the classroom or potential as source material for a mathematics club presentation.

THE AREA UNDER A PARABOLA BY CAVALIERI'S RULE

By **LYLE E. PURSELL**[*]

Grinnell College
Grinnell, Iowa

CAVALIERI'S rule for volumes ("Let K_1 and K_2 be two solids and let π_0 be some fixed reference plane. If for each plane π parallel to π_0 the area measures of the two plane sections $(K_1 \cap \pi)$ and $(K_2 \cap \pi)$ are

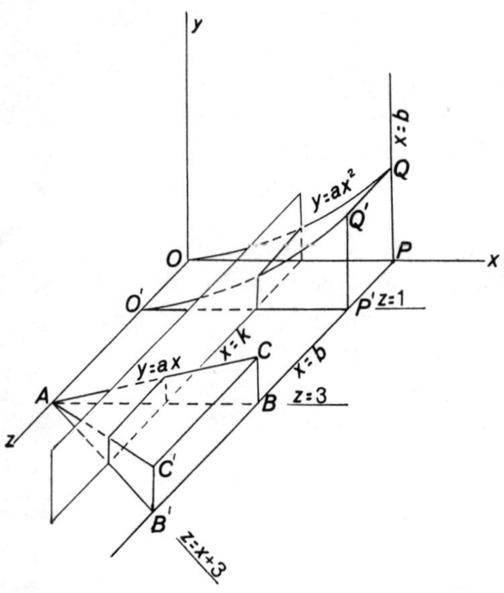

FIGURE 1

equal, then the two solids K_1 and K_2 have equal volume measures") is commonly used to find the volume measure of various solids.

---

[*] The author expresses his indebtedness to the referees for several suggestions for improving this paper.

In this paper we will show how this rule can also be used to find the area measure of the closed region bounded by the parabola $y = ax^2$, the x-axis, the origin, and the line $x = b$—assuming the area measure of this set exists—without explicitly using the integral calculus or other limit techniques. In Figure 1, OPQ is the closed, parabolic region

$$0 \le y \le ax^2, \qquad 0 \le x \le b$$

in the plane $z = 0$ whose area measure we wish to determine. We have constructed two auxiliary figures; $OPQO'P'Q'$ is the parabolic cylindrical solid with altitude parallel to the z-axis of length 1, while $ABCB'C'$ is the pyramid bounded by the planes $y = 0$, $y = ax$, $z = 3$, $z = x + 3$, and $x = b$. By a standard formula, the volume measure of the pyramid is given by

$$(1) \quad \text{Volume } (ABCB'C')$$

$$= \left(\frac{1}{3}\right) \cdot |AB| \cdot |BB'| \cdot |BC|$$

$$= \left(\frac{1}{3}\right)(b)(b)(ab)$$

$$= \frac{ab^3}{3}.$$

We will now use Cavalieri's rule to prove

$$(2) \quad \text{Volume } (ABCB'C') = \text{Volume } (OPQO'P'Q')$$
$$= \text{Area } (OPQ).$$

To prove the first equation in (2), consider an arbitrary plane $x = k$ parallel to the reference plane $x = 0$. If $k < 0$ or $b < k$, both intersections of the plane $x = k$ with the two solids are the empty set and have area measure 0. If $0 \le k \le b$, the intersection of this plane with the pyramid $ABCB'C'$ is the closed rectangular region

$$0 \leq y \leq ak, \qquad 3 \leq z \leq k + 3$$

of area measure $(ak)(k) = ak^2$, while its intersection with the parabolic cylindrical solid $OPQO'P'Q'$ is the rectangle

$$0 \leq y \leq ak^2, \qquad 0 \leq z \leq 1$$

of area measure

$$(ak^2)(1) = ak^2.$$

Since the two rectangular cross sections have the same area measure for all k such that $0 \leq k \leq b$, then by Cavalieri's rule

Volume $(ABCB'C')$ = Volume $(OPQO'P'Q')$.

The second equation of (2) follows from a standard formula for the volume of a cylindrical solid (which may also be proved using Cavalieri's rule). Comparing (1) and (2) we have

$$(3) \qquad \text{Area } (OPQ) = \frac{ab^3}{3}.$$

The reader should observe that we have not entirely avoided limits. Although Cavalieri's rule has a strong intuitive appeal, its rigorous proof depends on the limit process.

AREAS BY INFINITE SERIES

By LAWRENCE A. MORGAN

Montgomery County Community College
Blue Bell, Pennsylvania

ARCHIMEDES discovered a way of finding the exact area of plane surfaces bounded by curved lines without integral calculus.[1] The method consists of fitting simple figures like squares and triangles into the curve. The spaces between the curve and the simple figures are gradually filled up with smaller and smaller figures. If the areas of the simple figures can be calculated along with their infinite sum, the desired result is obtained. Two examples follow.

Parabola

To find the area between the parabola $y^2 = x$, the x-axis, and the line $x = \ell$, fill the curve with triangles as shown in figure 1. The first triangle drawn is right triangle OCB_0 with area $\frac{1}{2} \ell\sqrt{\ell}$. The hypotenuse of this triangle forms a side of the next triangle (vertical shading). The area of this triangle can be calculated by adding the areas of two triangles with base A_1B_1 and height $\sqrt{\ell}/2$. The y-coordinate of point A_1 is $\sqrt{\ell}/2$, while its x-coordinate is $\ell/4$. Point B_1 is the midpoint of the line OB_0, with coordinates $(\ell/2, \sqrt{\ell}/2)$. The area of triangle OA_1B_0 becomes $2[\frac{1}{2}(\ell/2 - \ell/4)(\sqrt{\ell}/2)] = \frac{1}{8}\ell\sqrt{\ell}$. There are now two spaces to be filled by triangles (horizontal shading). The area of triangle OA_2A_1 can be calculated by adding the areas of two triangles with base A_2B_2 and height $\sqrt{\ell}/4$. The y-coordinate of A_2 is $\sqrt{\ell}/4$, and its x-coordinate is $\ell/16$. Point B_2 is the midpoint of line $\overline{OA_1}$ with coordinates $(\ell/8, \sqrt{\ell}/4)$. The area of triangle OA_2A_1 becomes $2[\frac{1}{2}(\ell/8 - \ell/16)\sqrt{\ell}/4] = \frac{1}{64}\ell\sqrt{\ell}$; this can be shown in the same manner to be equal to the area of triangle $A_1A_2'B_0$, so that the total area of the horizontal shading is $\frac{1}{32}\ell\sqrt{\ell}$. This process continues, always placing one vertex of the smaller triangles on the curve (with the ordinate of the triangle closest to the x-axis half the ordinate of the previous triangle) and one side on a side of the triangles already drawn. The sum of the areas turns out to be the infinite geometric progression

$$\left(\frac{1}{2} + \frac{1}{8} + \frac{1}{32} + \frac{1}{128} + \cdots\right)\ell\sqrt{\ell}$$

of the type

$$a + ar + ar^2 + ar^3 + \cdots = \frac{a}{1 - r}$$

with

$$a = \tfrac{1}{2} \quad \text{and} \quad r = \tfrac{1}{4}.$$

Therefore the desired area is $\frac{2}{3}\ell\sqrt{\ell}$, or $\frac{2}{3}\ell^{3/2}$.

Circle

The area of a circle with radius r can be found by "filling" it up with triangles whose areas can be calculated (fig. 2). The first triangle is the large right triangle AOB with area $4(\frac{1}{2}rr) = 2r^2$. The four is necessary, since figure 2 shows only one-fourth of the desired area. The next triangle (ABD, vertical shading) is obtained by placing its base on the hypotenuse of the first triangle and placing its vertex on the curve in such a manner that the height of the triangle is part of a radius at $45°$ $\left(\theta_1 = \dfrac{\theta_0}{2} = \dfrac{90°}{2}\right)$ from the x-axis.

---

1. Archimedes' quadrature of the parabola is discussed in Eqmont Colerus' *From Simple Numbers to the Calculus,* published in 1955 by William Heinemann.

154

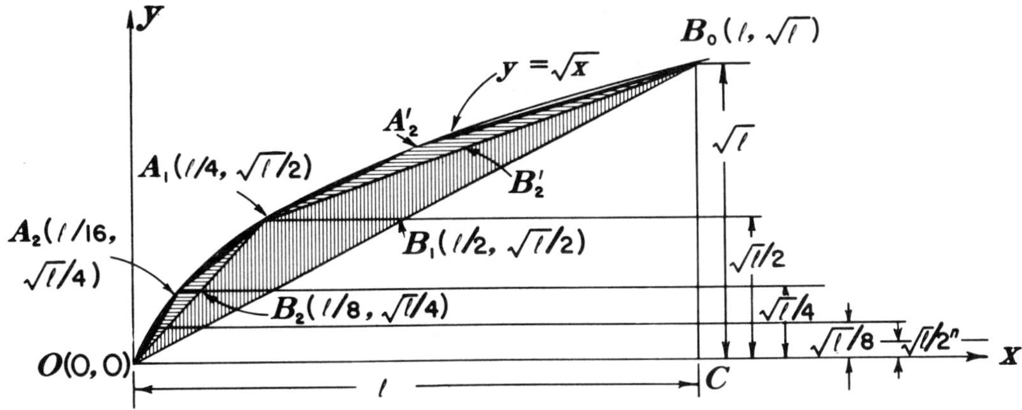

Fig. 1

The half-angle formulas

$$\cos \theta_n = \cos \frac{\theta_{n-1}}{2} = \sqrt{\frac{1 + \cos \theta_{n-1}}{2}}$$

and

$$\sin \theta_n = \sin \frac{\theta_{n-1}}{2} = \sqrt{\frac{1 - \cos \theta_{n-1}}{2}}$$

and the expressions $(r \cos \theta_n, r \sin \theta_n)$ for the coordinates (x_n, y_n) give for the point D the coordinates

(x_1, y_1)

$$= \left(r\sqrt{\frac{1 + \cos \theta_0}{2}} , r\sqrt{\frac{1 - \cos \theta_0}{2}} \right).$$

$$= (r\sqrt{2}/2, r\sqrt{2}/2)$$

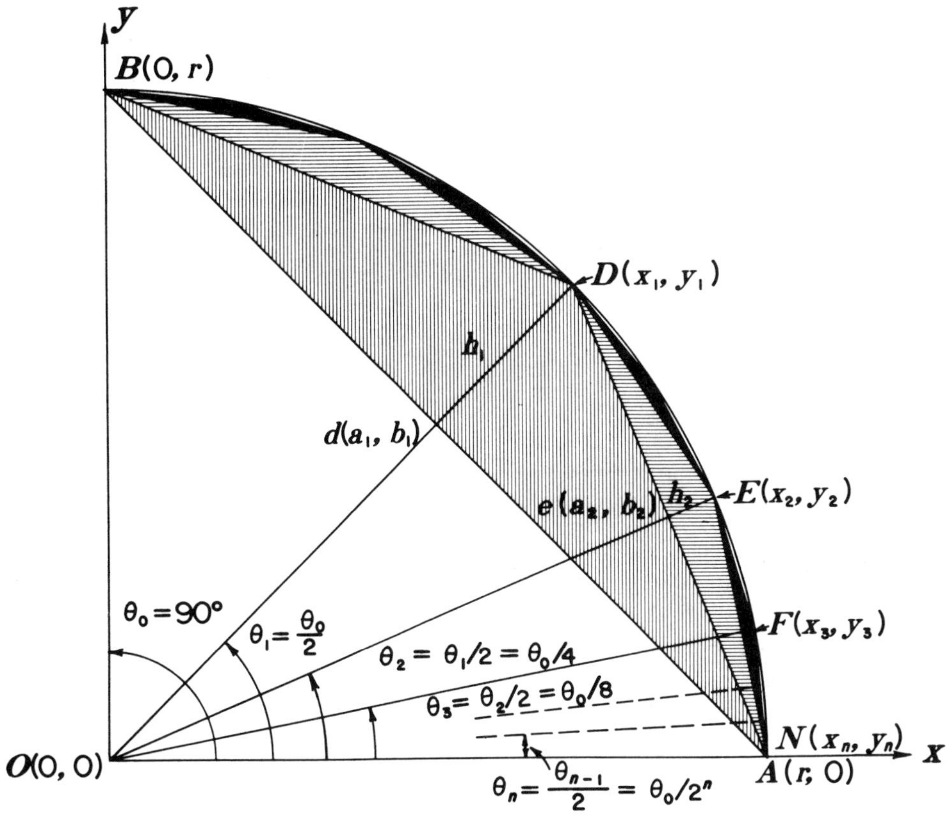

Fig. 2

TABLE 1
SERIES EXPANSION FOR π

i	θ_i	A_i (area)		$\sum_{i=0}^{i} A_i$	
0	$\theta_0 = 90°$	$2r^2$	$2.00r^2$	$= 2.00r^2$	
1	$\theta_1 = \theta_0/2$	$2r^2\sqrt{2} - 2r^2$	$2r^2\sqrt{2}$	$\approx 2.83r^2$	
2	$\theta_2 = \theta_1/2$	$2^2 r^2\sqrt{2 - \sqrt{2}} - 2r^2\sqrt{2}$	$2^2 r^2\sqrt{2 - \sqrt{2}}$	$\approx 3.06r^2$	
3	$\theta_3 = \theta_2/2$	$2^3 r^2\sqrt{2 - \sqrt{2 + \sqrt{2}}} - 2^2 r^2\sqrt{2 - \sqrt{2}}$	$2^3 r^2\sqrt{2 - \sqrt{2 + \sqrt{2}}}$	$\approx 3.12r^2$	
4	.	.	$2^4 r^2\sqrt{2 - \sqrt{2 + \sqrt{2 + \sqrt{2}}}}$	$\approx 3.14r^2$	
$\vdots$	$\vdots$	$\vdots$	$\vdots$		
n	$\theta_n = \theta_{n-1}/2$	$2^n r^2\sqrt{2 - S_n} - 2^{n-1} r^2\sqrt{2 - S_{n-1}}$	$\lim_{n \to \infty} 2^n r^2\sqrt{2 - S_n}$	$= \pi r^2$	

Since the coordinates of points A and B are known, the length of $\overline{AB}$ (the base of the triangle) and the coordinates of the midpoint $d(a_1, b_1)$ can be calculated. With coordinates d and D known, the height of the triangle is obtained (h_1). The area of the four triangles of this type in the circle becomes

$$4[\tfrac{1}{2}(\overline{AB})(h_1)]$$
$$= 4[\tfrac{1}{2}(\sqrt{(r - 0)^2 + (r - 0)^2})$$
$$\cdot \sqrt{(x_1 - a_1)^2 + (y_1 - b_1)^2}]$$
$$= 2r^2\sqrt{2} - 2r^2.$$

The next two triangles have identical areas (horizontal shading) and are obtained by placing their bases along the sides of the previous triangle. The vertex E of the triangle closest to the x-axis lies on the curve and along a radius at

$$22\tfrac{1}{2}° \left(\theta_2 = \frac{\theta_1}{2} = \frac{45°}{2} \right)$$

from the x-axis. The half-angle formulas are again used to obtain the coordinates of E:

$$(x_2, y_2) = \left(r\sqrt{\frac{1 + \cos \theta_1}{2}}, \ r\sqrt{\frac{1 - \cos \theta_1}{2}} \right)$$
$$= \left(r\sqrt{\frac{1 + \sqrt{2}/2}{2}}, \ r\sqrt{\frac{1 - \sqrt{2}/2}{2}} \right)$$
$$= (\tfrac{1}{2}r\sqrt{2 + \sqrt{2}}, \ \tfrac{1}{2}r\sqrt{(2 - \sqrt{2})}).$$

In a similar fashion we have a point for F:

$$(x_3, y_3) =$$
$$(\tfrac{1}{2}r\sqrt{2 + \sqrt{2 + \sqrt{2}}}, \ \tfrac{1}{2}r\sqrt{2 - \sqrt{2 + \sqrt{2}}}).$$

The area of the eight triangles of this type in the circle becomes

$$8[\tfrac{1}{2}(\overline{AD})h_2] = 8[\tfrac{1}{2}(\sqrt{(r - x_1)^2 + y_1^2})$$
$$\cdot (\sqrt{(x_2 - a_2)^2 + (y_2 - b_2)^2})]$$
$$= 4r^2\sqrt{2 - \sqrt{2}} - 2r^2\sqrt{2},$$

with the area of the sixteen triangles of this type equal to

$$8r^2\sqrt{2 - \sqrt{2 + \sqrt{2}}} - 4r^2\sqrt{2 - \sqrt{2}}.$$

The derivation continues in this manner, resulting in the unusual series expansion (or radical-chain) for π shown in table 1. where

$$S_1 = 0$$
$$S_2 = \sqrt{2 + S_1}$$
$$S_3 = \sqrt{2 + S_2}$$
$$\vdots$$
$$S_n = \sqrt{2 + S_{n-1}}$$

One could consider other curves (an ellipse might be the next logical extension) with this exercise in coordinate geometry, trigonometry, induction, series, and limits. If nothing else, this should give one a greater appreciation of the calculus.

Bibliography: *Integration—Pedagogy*

Reynolds, Joseph B. Finding Plane Areas by Algebra. 21 (April 1928): 197–203.
 Remarks on calculating plane areas using series summation.

Boyer, Carl B. The Quadrature of the Parabola: An Ancient Theorem in Modern Form. 47 (January 1954): 36–37.
 Some remarks on another approach to the squaring of the parabola.

Yates, Robert C. The Cardioid. 52 (January 1959): 10–14.
 Remarks on the cardioid as a roulette, conchoid, pedal, caustic, and envelope. Also includes a determination of its arc length and area.

McIntosh, Jerry A. Determining the Area of a Parabola. 66 (January 1973): 88–91.
 Remarks on finding the area bounded by a parabola and an intersecting straight line.

Ocrant, Ian. A Guide to Mathematical Discovery. 66 (April 1973): 331–35.
 Remarks from a student's standpoint concerning the discovery method at work. Includes a BASIC program for calculating areas.

ON THE EVALUATION OF $\int \sec x\, dx$ AND $\int \csc x\, dx$

M. J. HELLMAN

Rutgers—The State University
Newark, New Jersey

WHEN the integration of sec x and csc x is first attempted in the elementary calculus course it becomes clear that these quantities are not the end products of differentiating any simple functions the student has yet encountered. It appears to the author that the best technique and one which is most naturally motivated is to examine the simple trigonometric identities already encountered to see whether any can be found which relate sec x and csc x to functions which are more readily integrated, that is to functions which are recognizable as end products of differentiating elementary functions. Very quickly these elementary identities are found:

$$(1) \qquad \sec x - \tan x = \frac{\cos x}{1 + \sin x}$$

$$(2) \qquad \csc x - \cot x = \frac{\sin x}{1 + \cos x}$$

Using (1) it follows that

$$(3) \quad \int \sec x\, dx$$

$$= \int \tan x\, dx + \int \frac{\cos x\, dx}{1 + \sin x}$$

$$= -\ln |\cos x| + \ln |1 + \sin x| + C$$

$$= \ln \left| \frac{1 + \sin x}{\cos x} \right| + C$$

$$= \ln |\sec x + \tan x| + C$$

Similarly, from (2) it follows that

$$(4) \quad \int \csc x\, dx$$

$$= \int \cot x\, dx + \int \frac{\sin x\, dx}{1 + \cos x}$$

$$= \ln |\sin x| - \ln |1 + \cos x| + C$$

$$= \ln \left| \frac{\sin x}{1 + \cos x} \right| + C$$

$$= \ln \left| \frac{1 - \cos x}{\sin x} \right| + C$$

$$= \ln |\csc x - \cot x| + C$$

Once having obtained (3), if it is desired to avoid the parallel argument to obtain (4), the latter can be derived alternatively as follows:

$$(5) \quad \int \csc x\, dx$$

$$= \int \sec \left(\frac{\pi}{2} - x \right) dx$$

$$= -\ln \left| \sec \left(\frac{\pi}{2} - x \right) + \tan \left(\frac{\pi}{2} - x \right) \right| + C$$

$$= \ln \left| \frac{1}{\sec \left(\frac{\pi}{2} - x \right) + \tan \left(\frac{\pi}{2} - x \right)} \right| + C$$

$$= \ln \left| \sec \left(\frac{\pi}{2} - x \right) - \tan \left(\frac{\pi}{2} - x \right) \right| + C$$

$$= \ln |\csc x - \cot x| + C$$

AN ALGORITHM for PARTIAL-FRACTION EXPANSION

By JOSEPH W. ROGERS

Bucknell University
Lewisburg, Pennsylvania

and MARGARET ANNE ROGERS

Susquehanna University
Selinsgrove, Pennsylvania

WE USUALLY expand a rational function by first reducing it to the sum of a polynomial and a proper fraction that is the quotient of two polynomials in which the degree of the numerator is less than the degree of the denominator. We expand the fraction by writing it as a sum of partial fractions with undetermined numerator coefficients. For example,

$$\frac{8x^3+28x^2+28x+17}{(x+2)^3\cdot(x-1)^2}=\frac{a_3}{(x+2)^3}+\frac{a_2}{(x+2)^2}$$

$$+\frac{a_1}{(x+2)}+\frac{b_2}{(x-1)^2}+\frac{b_1}{(x-1)}.$$

There are two standard techniques for evaluating the coefficients.

One technique consists of recombining the partial-fraction expansion into a single fraction and equating like powers of its numerator polynomial with those of the numerator polynomial of the original fraction. If the number of undetermined coefficients is n, there would be n linearly independent equations in n unknowns. These could be solved to find any of the coefficients. In the above example, the following would have to be solved:

$$a_1 + b_1 = 0$$

$$2a_1 + a_2 + 5b_1 + b_2 = 8$$

$$-3a_1 + a_3 + 6b_1 + 6b_2 = 28$$

$$-4a_1 - 3a_2 - 2a_3 - 4b_1 + 12b_2 = 28$$

$$4a_1 + 2a_2 + a_3 - 8b_1 + 8b_2 = 17.$$

The second method consists of multiplying by the highest power of each linear factor of the denominator (one at a time), canceling as far as possible, and then finally setting the factor equal to zero. To find a_3 in the example, we multiply by $(x + 2)^3$ and obtain

$$\frac{8x^3 + 28x^2 + 28x + 17}{(x - 1)^2} = a_3.$$

Since $x = -2$, a_3 is found to be 1.

The coefficients that correspond to the greatest powers of the respective denominator factors can be found. In the above example, a_3 and b_2 can be calculated in this manner. All the remaining coefficients are then computed using the first method. In this example, there would still be three linearly independent equations in three unknowns.

The task of solving a large number of equations can be tedious. Furthermore, all the unknowns must be involved even if only one is being sought. The purpose of this paper is to present an algorithm that reduces the amount of required calculations considerably. We shall also be able to find any one of the numerator coefficients without always involving all the others. (As an added extra, the procedure is easily adapted for solution on a digital computer.)

Our development begins with a few manipulations in order to put the given fraction into useful form. After identifying some of the unknowns of the partial-fraction expansion, we shall deduce the algorithm.

Let us consider $F(x)$ as a proper fraction whose denominator, polynomial $Q_0(x)$, has a zero z with multiplicity k. Then $F(x)$ can be written as

$$F(x) = \frac{P_0(x)}{Q_0(x)} = \frac{P_0(x)}{Q_1(x)(x-z)^k}$$

where $Q_1(x)$ is a polynomial that does not contain the factor $(x - z)$. We could expand to get

$$F(x) = \frac{a_k}{(x-z)^k} + \frac{a_{k-1}}{(x-z)^{k-1}}$$

$$+ \cdots + \frac{a_1}{(x-z)} + \frac{Q_2(x)}{Q_1(x)}.$$

$Q_2(x)/Q_1(x)$ is also a proper fraction that we could expand to get a complete partial-fraction expansion for $F(x)$. We do not need to do this right now, however. Upon recombining we get

$$F(x) =$$

$$\frac{Q_1(x)[a_k+a_{k-1}(x-z)+\cdots+a_1(x-z)^{k-1}]}{Q_1(x)(x-z)^k}$$

$$+\frac{(x-z)^kQ_2(x)}{Q_1(x)(x-z)^k}.$$

So far we have really accomplished nothing except having powers of $(x - z)$ appropriately distributed in the numerator polynomial. The coefficient a_j is associated with $(x - z)^{k-j}$ for each integer j between and including 1 and k. In addition, $Q_2(x)$ is multiplied by $(x - z)^k$. This distribution of powers of $(x - z)$ is the key to our algorithm.

Let us observe what happens when x is equal to z. First of all, $P_0(z) = a_k Q_1(z)$, since every other term in the numerator contains the factor $(x - z)$, which is now zero. Hence, we easily find a_k to be $P_0(z)/Q_1(z)$. Secondly, we see that $P_0(z) - a_k Q_1(z)$ is zero, so that $(x - z)$ is a factor of $P_0(x) - a_k Q_1(x)$. We obtain a new polynomial $P_1(x)$ by removing this factor. That is,

$$P_1(x)(x - z) = P_0(x) - a_kQ_1(x).$$

We see then that $P_1(z) = a_{k-1} Q_1(z)$. Therefore, $a_{k-1} = P_1(z)/Q_1(z)$. Also, since $P_1(z) - a_{k-1} Q_1(z) = 0$, then $(x - z)$ is a factor of $P_1(x) - a_{k-1} Q_1(x)$. Following the form of the definition for $P_1(x)$, we

define $P_2(x)$ as the polynomial that results from removing the factor $(x - z)$ from $P_1(x) - a_{k-1} Q_1(x)$. That is,

$$P_2(x)(x - z) = P_1(x) - a_{k-1}Q_1(x).$$

We follow this procedure until we find $a_k, a_{k-1}, \cdots, a_1$. Each new polynomial is defined recursively through

$$P_m(x)(x - z) = P_{m-1}(x) - a_{k-m+1}Q_1(x),$$

from which it follows that $a_{k-m} = P_m(z)/Q_1(z)$. As a by-product we have $P_k(x) = Q_2(x)$. This last result becomes clear as we reexamine the whole procedure. Effectively at the first step, we are forming $F(x) - \frac{a_k}{(x-z)^k}$. Let us designate this as $F_1(x)$ so that we have a natural recursive formula

$$F_m(x) = F_{m-1}(x) - \frac{a_{k-m+1}}{(x-z)^{k-m+1}}$$

from which it follows that

$$F_k(x) = F_{k-1}(x) - \frac{a_1}{(x-z)} = \frac{Q_2(x)}{Q_1(x)}.$$

We can now deduce the algorithm quite easily. We extract the following plan from the foregoing manipulations.

1. We factor the denominator polynomial, identifying one of the factors as $(x - z)$, and then determine the multiplicity k of this factor.

2. We identify $Q_1(x)$ as the polynomial that results from factoring $(x - z)^k$ from $Q_0(x)$.

3. We compute $P_0(z)$ and $a_k = P_0(z)/Q_1(z)$ and next determine $P_1(x)$ from $P_1(x) (x - z) = P_0(x) - a_k Q_1(x)$.

4. Repeating the previous step as many times as necessary, we find $a_{k-1}, a_{k-2}, \cdots, a_1$ and $P_2(x), \cdots, P_k(x)$ from $a_m = P_{k-m}(z)/Q_1(z)$ and $P_m(x) (x - z) = P_{m-1}(x) - a_{k-m+1} Q_1(x)$.

5. Recognizing that $P_k(x) = Q_2(x)$, we have a new fraction $Q_2(x)/Q_1(x)$, which can be further expanded into its partial-fraction expansion by starting again at step 1 above. Eventually we can repeat

the procedure to get, if necessary, the complete partial-fraction expansion for $F(x)$.

We offer the following example in order to illustrate the computational technique. It is the same problem we presented at the beginning of this paper, and the designations for the numerator coefficients will be the same.

$$F(x) = \frac{8x^3 + 28x^2 + 28x + 17}{(x+2)^3 \cdot (x-1)^2}.$$

We start by identifying z as -2. Consequently, $k = 3$; $Q_1(x)$ is $(x-1)^2$; and $P_0(x)$ is $8x^3 + 28x^2 + 28x + 17$.

$$a_3 = \frac{P_0(-2)}{Q_1(-2)} = \frac{9}{9} = 1.$$

$$P_1(x)(x+2) = P_0(x) - a_3 Q_1(x)$$
$$= 8x^3 + 27x^2 + 30x + 16,$$

so that

$$P_1(x) = 8x^2 + 11x + 8.$$

$$a_2 = \frac{P_1(-2)}{Q_1(-2)} = \frac{18}{9} = 2.$$

$$P_2(x)(x+2) = P_1(x) - a_2 Q_1(x)$$
$$= 6x^2 + 15x + 6,$$

so that

$$P_2(x) = 6x + 3.$$

$$a_1 = \frac{P_2(-2)}{Q_1(-2)} = \frac{-9}{9} = -1.$$

$$P_3(x)(x+2) = P_2(x) - a_1 Q_1(x)$$
$$= x^2 + 4x + 4,$$

so that

$$P_3(x) = x + 2 = Q_2(x).$$

We are left with $(x+2)/(x-1)^2$ to analyze. (In order to avoid confusion, let us use the notations z', k', $P_0'(x)$, $P_1'(x)$, $P_2'(x)$, and $Q_1'(x)$ instead of their unprimed counterparts and b_2 and b_1 instead of a_2 and a_1 as designated in the algorithm.) $z' = 1$, and $k' = 2$. $P_0'(x) = x + 2$, and $Q_1'(x)$ is 1 (a constant polynomial).

$$b_2 = \frac{P_0'(1)}{Q_1'(1)} = \frac{3}{1} = 3.$$

$$P_1'(x)(x-1) = P_0'(x)$$
$$- b_2 Q_1'(x) = x - 1,$$

so that $P_1'(x) = 1$ (a constant polynomial).

$$b_1 = \frac{P_1'(1)}{Q_1'(1)} = \frac{1}{1} = 1.$$

$$P_2'(x) = P_1'(x) - b_1 Q_1'(x) = 0$$

(the zero polynomial). At this point we realize that we have completed the expansion:

$$\frac{8x^3+28x^2+28x+17}{(x+2)^3 \cdot (x-1)^2} = \frac{1}{(x+2)^3} + \frac{2}{(x+2)^2}$$
$$+ \frac{-1}{(x+2)} + \frac{3}{(x-1)^2} + \frac{1}{(x-1)}.$$

We see that this algorithm requires much less computing than the usual methods. There are no large sets of equations to solve. We have only a sequence of simple calculations that lead to results quite rapidly. If the linear factors of the denominator all have multiplicity 1, then this algorithm has little advantage, if any, over the second one listed at the beginning of this paper. However, when the multiplicity exceeds 1, it has an advantage that becomes increasingly greater as the multiplicity of the factors increases. This algorithm is clearly superior to the first approach. It can easily be adapted so that the coefficients can be calculated on a digital computer. One disadvantage is that the method is not intuitively obvious.

Bibliography: *Techniques of Integration, Special Integrals*

Smith, Eugene R. Decomposition into Partial Fractions by Means of Remainders. 6 (June 1914): 203–8.
  Remarks on simplifying the computational procedure in partial fraction decomposition.

Metzler, George F. Decomposition into Partial Fractions. 7 (June 1915): 159–66.
  Remarks on simplifying the computational procedure in partial fraction decomposition.

Smith, Eugene R. Decomposition into Partial Fractions. 8 (March 1916): 132–44.
  Remarks on simplifying the computational procedure in partial fraction decomposition.

Court, Nathan Altshiller. Mascheroni Constructions. 51 (May 1958): 370–72.
  Remarks on Mascheroni's contribution to calculus and geometry.

Just, Erwin, and Norman Schaumberger. Two Theorems Concerning Partial Fractions. 59 (January 1966): 34–35.
  Some remarks on partial fraction decomposition.

Baylock, Adrian F. Letter to the editor. 61 (October 1968): 641.
  Some curious integrals.

Trigg, Charles W. Letter to the editor. 63 (December 1970): 645.
  Some curious integrals.

Schaumberger, Norman. The Evaluation of $\int_a^b \frac{dx}{x^2}$ and $\int_a^b \frac{dx}{\sqrt{x}}$ 64 (November 1971): 605–6.

Iacobacci, Rora. On Stieltjes Integrals. 65 (May 1972): 479–85.
  A discussion of Stieltjes integrals with examples of their use.

MOMENT OF INERTIA Problem—
A Classroom PARADOX

By PHILIP J. LEHPAMER

St. Dominic College
St. Charles, Illinois

I RECENTLY assigned the following calculus problem to my class: Find the moment of inertia of a cone of radius r and altitude h about its axis, if its mass density is D. This problem caused considerable controversy among students and stimulated the entire class to discussion, for two different methods of solution had apparently produced two different answers.

One student set the problem up as in figure 1 and generated a cone by revolving the triangle bounded by the axes and the line $y = h(r - x)/r$ about the y-axis. The

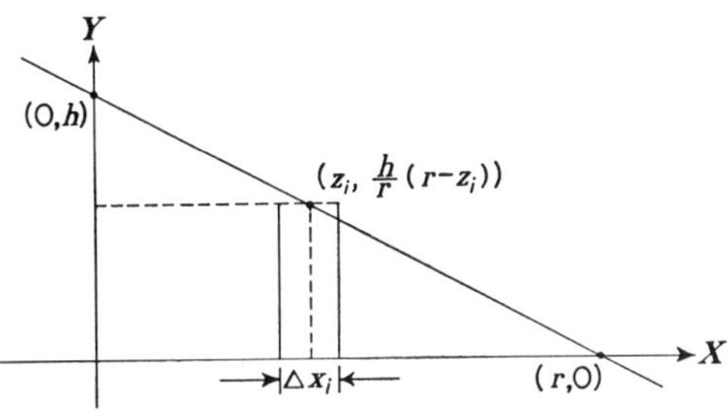

FIGURE 1

student partitioned the x-axis and examined the ith rectangle. Revolving this rectangle about the y-axis, a pipe-shaped figure is obtained whose volume is $2\pi z_i h(r - z_i)\Delta x_i/r$ and whose mass is $2\pi D z_i h(r - z_i)\Delta x_i/r$. If the mass of the pipe were all concentrated at the distance z_i from the axis, the moment of inertia of the pipe would be

$$(1) \qquad 2\pi D z_i h(r - z_i)\Delta x_i z_i^2/r,$$

and this approximates the actual moment of inertia of the pipe. Considering all the pipes generated by the rectangles, summing and taking the limit yields

$$(2) \qquad \frac{2\pi Dh}{r} \int_0^r x^3(r - x)dx.$$

Whereupon, the fundamental theorem gives $\pi Dhr^4/10$ as the moment of inertia of the cone about its axis.

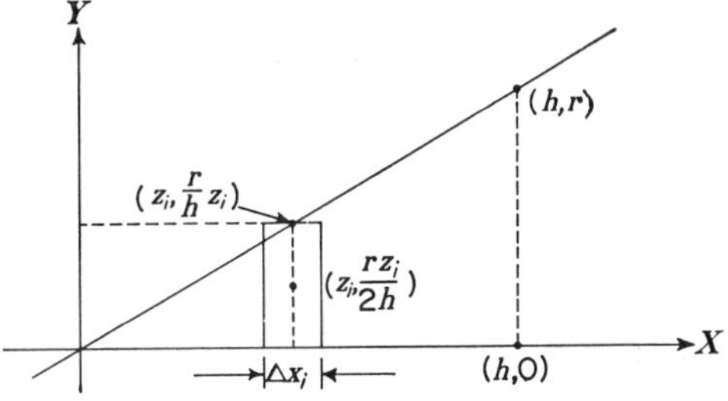

FIGURE 2

A second student set the problem up as in figure 2 and generated a cone by revolving the triangle bounded by $x = h$, $y = 0$, and $y = rx/h$ about the x-axis. This student also partitioned the x-axis and examined the ith rectangle. Revolving this rectangle about the x-axis, a cylinder is obtained whose volume is $\pi(rz_i/h)^2\Delta x_i$ and whose mass is $\pi D(rz_i/h)^2\Delta x_i$. Then the student argued that since the center of mass of a uniform rectangle is at its geometric center, the moment of inertia of the ith figure is approximated by

$$(3) \qquad \pi D(rz_i/h)^2\Delta x_i(rz_i/2h)^2.$$

Summing and taking the limit yields

$$(4) \qquad \frac{\pi Dr^4}{4h^4} \int_0^h x^4 dx = \pi Dhr^4/20$$

as the moment of inertia of the cone about its axis.

At first, the class thought they had discovered a mathematical paradox. Then they divided evenly over which of the two solutions was correct. Finally, after discussion, the group supporting the first student convinced the other group that the second solution had a flaw. They argued that expression (3) was incorrect since it involved the mass of a *solid* (i.e., $\pi D(rz_i/h)^2\Delta x_i$) and the distance to the axis of the center of mass of a *plane* figure (i.e., $(rz_i/2h)^2$). In order to remedy the situation, the class decided that figure 2 should be partitioned along the y-axis and horizontal rectangles used rather than vertical rectangles. This reduced the problem to the approach used in the first solution.

Finally, a student suggested that since we had previously calculated the moment of inertia of a solid cylinder about its axis (namely, $\pi Dhr^4/2$), this result could be used with figure 2. Thus, the moment of inertia of the cone is

$$(5) \qquad \frac{\pi Dr^4}{2h^4}\int_0^h x^4 dx = \pi Dhr^4/10,$$

since the moment of inertia of the ith cylinder is

$$(6) \qquad \pi D(\Delta x_i)(rz_i/h)^4/2.$$

This is another example of a situation in which a mistake in reasoning can lead to exciting dialogue and productive learning in mathematics.

THE SCHWARZ PARADOX: An INTERESTING PROBLEM for the FIRST-YEAR CALCULUS STUDENT

By GAIL H. ATNEOSEN

Western Washington State College
Bellingham, Washington

THE following result is well known in that it frequently appears tucked away in a problem list at the end of a beginning calculus text. Thus, unfortunately, it is too often omitted or overlooked. However, a solution is within the grasp of a first-year student of calculus. The problem challenges the student to use elementary notions from trigonometry and plane geometry plus the newly acquired concept of limit to obtain a surprising and important mathematical result.

The German mathematician H. A. Schwarz (1890, p. 309) established that it is possible to approximate the lateral surface of a right circular cylinder of finite dimensions by an inscribed polyhedral surface of arbitrarily large area. This surprising result is sometimes known as the "Schwarz paradox." Schwarz's result establishes that the type of approximation used in the interpretation of the definite integral as the area under a curve does not easily generalize to the area of curved surfaces. The purpose of this note is to bring attention to this result as an interesting problem for the student.

Solution to the Problem

The curved surface we wish to approximate is the lateral surface of a right cir-cular cylinder of height 1 with base a circle whose radius is 1. In order to describe the polyhedral surface that is to approximate the cylinder, the cylinder is cut along a generator and then spread out on the plane to obtain a rectangle with sides of lengths 1 and 2π. The side of length 1 is divided into m equal parts and the side of length 2π into n equal parts; the rectangle is divided into mn smaller rectangles by drawing lines parallel to the sides through the points of subdivision. Figure 1 shows a subdivision for $m = 3$ and $n = 6$. Next, each of the mn rectangles is divided into 4 triangles by drawing diagonals as pictured in figure 2. Then the rectangle is rolled back up into a cylinder. Each of the curved triangles is replaced by a flat triangle with the same vertices, as shown in figure 3. The collection of these $4mn$ flat triangles constitutes the polyhedral surface that approximates the cylinder.

Next the area of this polyhedral surface must be calculated. A typical curved rectangle and the four flat triangles are shown in figure 4. The points A, B, C, D, H, P, and S lie on the surface of the cylinder. The inscribed polyhedral surface consists of $2mn$ flat triangles congruent to triangle PBD and $2mn$ flat triangles congruent to triangle CPD.

The area of triangle $PBD = \dfrac{1}{2} DB \cdot PH$.

Observe that $PH = SD$, and by the

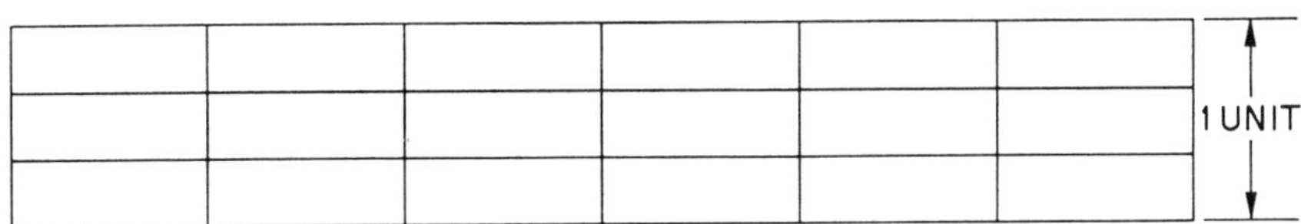

Fig. 1

165

"

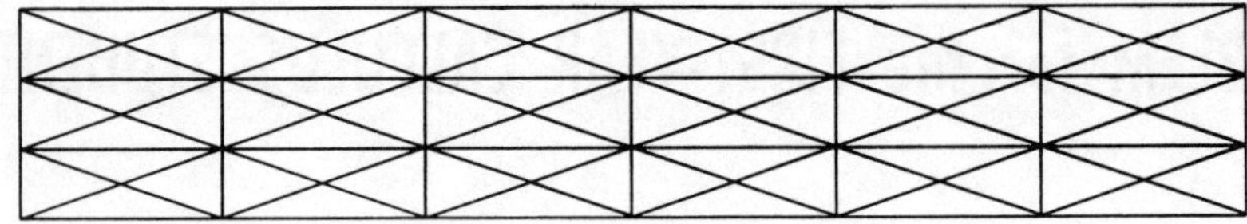

Fig. 2

Pythagorean theorem

$$SD = [(FS)^2 + (FD)^2]^{1/2}$$

where

$$FS = OS - OF = 1 - \cos\frac{\pi}{n} \quad \text{and} \quad FD = \sin\frac{\pi}{n}.$$

Hence

$$SD = \left[\left(1 - \cos\frac{\pi}{n}\right)^2 + \left(\sin\frac{\pi}{n}\right)^2\right]^{1/2}.$$

Squaring the first term and using the trigonometric identities

$$\sin^2\left(\frac{\pi}{n}\right) + \cos^2\left(\frac{\pi}{n}\right) = 1$$

and

$$2\sin^2\left(\frac{\pi}{2n}\right) = 1 - \cos\frac{\pi}{n}$$

yields

$$SD = 2\sin\frac{\pi}{2n}.$$

Since

$$DB = \frac{1}{m} \quad \text{and} \quad PH = 2\sin\frac{\pi}{2n},$$

the area of triangle $PBD = \dfrac{1}{m}\sin\dfrac{\pi}{2n}$.

The area of triangle $CPD = \dfrac{1}{2}\,CD \cdot FP.$ By the Pythagorean theorem,

$$FP = [(SP)^2 + (FS)^2]^{1/2}.$$

Since $SP = \dfrac{1}{2m}$ and $FS = 1 - \cos\dfrac{\pi}{n}$, as noted above, it follows that

$$FP = \left[\left(\frac{1}{2m}\right)^2 + \left(1 - \cos\frac{\pi}{n}\right)^2\right]^{1/2}.$$

Utilizing the second of the above trigonometric identities yields

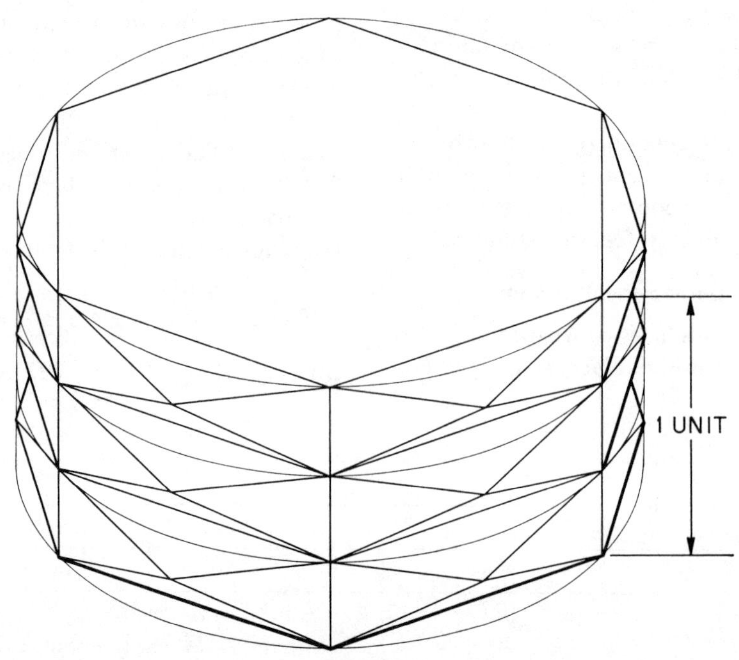

Fig. 3

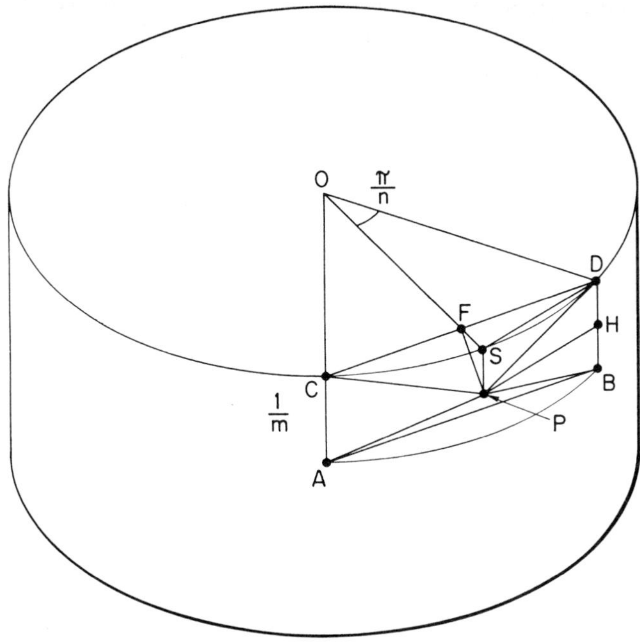

Fig. 4

$$FP = \left[\frac{1}{4m^2} + 4\sin^4\left(\frac{\pi}{2n}\right)\right]^{1/2}.$$

Since $\frac{1}{2} CD = FD$ and $FD = \sin\frac{\pi}{n}$, it follows that the area of triangle

$$CPD = \left[\sin\frac{\pi}{n}\right]\left[\frac{1}{4m^2} + 4\sin^4\left(\frac{\pi}{2n}\right)\right]^{1/2}.$$

The total area A of the inscribed polyhedral surface is:

$$A = 2mn\left[\frac{1}{m}\sin\frac{\pi}{2n}\right] + 2mn\left[\sin\frac{\pi}{n}\right]\left[\frac{1}{4m^2} + 4\sin^4\left(\frac{\pi}{2n}\right)\right]^{1/2}.$$

What happens to the area of the inscribed polyhedral surface as m and n get large can be determined more readily if we express A as follows:

$$A = \pi\left(\frac{2n}{\pi}\sin\frac{\pi}{2n}\right)$$

$$+ \pi\left(\frac{n}{\pi}\sin\frac{\pi}{n}\right)\left[1 + \frac{\pi^4 m^2}{n^4}\left(\frac{2n}{\pi}\sin\frac{\pi}{2n}\right)^4\right]^{1/2}.$$

Here we need the result that

$$\lim_{x\to 0}\frac{1}{x}\sin x = 1.$$

Thus, as n becomes large,

$$\frac{2n}{\pi}\sin\frac{\pi}{2n} \qquad \text{and} \qquad \frac{n}{\pi}\sin\frac{\pi}{n}$$

approach 1. In order to approximate A, we can replace these quantities by 1, yielding

$$A \sim \pi + \pi\left(1 + \frac{\pi^4 m^2}{n^4}\right)^{1/2}$$

If $m = n^3$, then

$$A \sim \pi[1 + (1 + \pi^4 n^2)^{1/2}].$$

Thus A can be made as large as we please by allowing n to get large and setting $m = n^3$. Furthermore, the larger the values of m and n, the better the inscribed polyhedral surface approximates the surface of the cylinder. (It is also interesting to consider what happens when m is not equal to n^3.)

This result could be presented to the student as a challenging problem (with some of the intermediate steps indicated), or it could be the basis for a lively classroom discussion. The result is of interest, and the techniques used to obtain the result are very instructive. The notion of limit is difficult for some students to grasp; this problem involves the student with a geometric situation in which to work with the notion of limits.

REFERENCES

Olmsted, John M. H. *Advanced Calculus.* New York: Appleton-Century-Crofts, 1961. Pp. 616–17.

Schwarz, H. A. *Gesammelte Mathematische Abhandlungen.* Vol. 2. Berlin: Julius Springer, 1890.

Thomas, George B. *Calculus and Analytic Geometry.* 4th ed. Reading, Mass.: Addison-Wesley Publishing Co., 1968. Pp. 568–70.

Bibliography: *Applications of Integration*

Court, Nathan Altshiller. Notes on the Centroid. 53 (January 1960): 33–35.

Historical remarks citing contributions of such mathematicians as Archimedes, Commandino, Maurolycus, Leonardo da Vinci, Monge, and Carnot.

Chow Chi-Ming. The Relation between Distance and Sight Area. 58 (April 1965): 298–302.

Some remarks concerning the size and shape of what we see.

Damaskos, Nickander J. A Case Study in Mathematics—the Cone Problem. 62 (December 1969): 642–49.

Remarks concerning the use of computers as a teaching aid in calculating the volumes of solids.

8

Numerical and Mechanical Methods

Through the years, many numerical and mechanical methods have been devised to handle the area and volume problems of integral calculus. Two of these are described here.

The article by Meserve and Pingry deals with the application of the prismoidal formula to the calculation of volumes. Kays discusses the use of an integraph for the solution of area problems. A step-by-step method for constructing this instrument is given.

This material can be used either directly in the classroom or in a mathematics club presentation. It can serve to broaden the perspective of students who, too often, get the impression that the method in the textbook is the only available one.

Some Notes on the Prismoidal Formula

By B. E. Meserve *and* R. E. Pingry
University of Illinois, Urbana, Illinois

Introduction

In this article we shall discuss the wide applicability of the prismoidal formula to the calculation of the volumes of many common solids, as well as many that are not so common. Some explicit criteria for the application of the formula will be presented. The writers believe that the prismoidal formula may serve as an excellent review of most of the volume relationships of solid geometry, as well as a review of all the geometric relationships necessary to obtain the area of the bases and midsections.

A *prismoid* is a special case of a prismatoid. A *prismatoid* is a polyhedron (a solid bounded by planes) all of whose vertices lie in two parallel planes. The faces of the polyhedron lying in the two parallel planes are called the *bases* of the prismatoid. If the two bases of the prismatoid have the same number of sides, the prismatoid is called a *prismoid*. [2]* If those edges of a prismoid not lying in the bases of the prismoid are parallel, the solid is called a *prism*. The prismoidal formula may be used for many solids including all prismatoids and therefore for all prismoids

---

The volumes of a

$$\left.\begin{array}{l}\text{Rectangular parallelepiped}\\ \text{Cylinder}\\ \text{Prism}\\ \text{Pyramid}\\ \text{Cone}\\ \text{Prismatoid}\\ \text{Prismoid}\\ \text{Sphere}\\ \text{Spherical segment}\\ \text{Spheroid}\\ \text{Conoid}\\ \text{Wedge}\\ \text{Ellipsoid, and many other solids.}\\ \text{(See Section B)}\end{array}\right\} = \frac{H}{6}\,(B_1+4M+B_2)$$

H = Altitude
B_1 = Area of first base
B_2 = Area of second base
M = Area of midsection

The *prismoidal formula,* $V = H/6(B_1 + 4M + B_2)$, is considered in many high school solid geometry textbooks in connection with a topic on the prismatoid. In these texts the formula is frequently called the prismatoid formula; however, in several calculus textbooks and books of tables this formula is called the prismoidal formula. The writers shall use the latter term in discussing the formula throughout this paper.

and prisms. Some high school textbooks only consider the application of the prismoidal formula to prismatoids. Other textbooks have exercises that demonstrate the use of the formula in finding volumes of more general solids. In particular, Frame's *Solid Geometry* [2] contains an excellent treatment of the subject matter of

* Numbers in brackets refer to references listed at the end of the paper.

this paper. This formula is applicable not only to most of the solids usually studied in a high school solid geometry course, but also to many solids having unusual shapes. For example, the prismoidal formula can be used to find the volumes of many of the solids appearing in the exercises of calculus textbooks.

When can the formula be used? When can it not be used? In an attempt to answer these questions, the remainder of the paper is divided into two sections. The first section (Section A) contains examples of the use of the prismoidal formula for most of the common solids, and a few solids that are frequently considered in calculus textbooks. The second section (Section B) contains some explicit criteria for the general applicability of the formula.

A. Application of the Prismoidal Formula to Common Solids

I. Solids for Which $V = BH$

We shall consider here solids that have two parallel bases. Any section of the solid by a plane parallel to these bases is called a *principal section*. If all the principal sections of a solid have equal areas, the volume of the solid is equal to the area of its base multiplied by its altitude. Such a solid might be illustrated by a uniform pack of cards stacked on a table in any manner which leaves the cards parallel to the table top. The common solids for which $V = BH$, however, can generally be described as being enclosed by a closed cylindrical or prismatic surface and two parallel planes that intersect all elements of this surface. These solids are called cylinders and prisms. For each of these, the application of the prismoidal formula is almost trivial since the two bases and the midsection are congruent, i.e., $B_1 = M = B_2$ and thus the prismoidal formula becomes

$$V = \frac{H}{6}(B_1 + 4M + B_2) = \frac{H}{6}(6B_1) = BH.$$

II. Solids for Which $V = BH/3$

These solids include right pyramids, right circular cones, and in general, all solids enclosed by a pyramidal or conical surface and a plane cutting all the elements of the surface. The prismoidal formula can be easily applied to these solids because the area of one base is zero (we shall take $B_2 = 0$) and the area of the midsection is always one-fourth of the area of the other base, i.e., $M = B_1/4$. For example, any circular cone may be considered as having $B_1 = \pi r^2$, $B_2 = 0$, $M = \pi(r/2)^2 = \pi r^2/4$, and height H.

$$V = \frac{H}{6}\left(\pi r^2 + \frac{4\pi r^2}{4} + 0\right) = \frac{\pi r^2 H}{3} = \frac{BH}{3}.$$

III. Solids for Which $V = BH/2$.

We shall briefly mention two types of solids, wedges and conoids, for which $V = BH/2$. A *wedge* may be defined as a triangular right prism with one face that is not a base of the prism taken as a base of the wedge. All principal sections are then rectangles such as $HIJK$ in Figure 1.

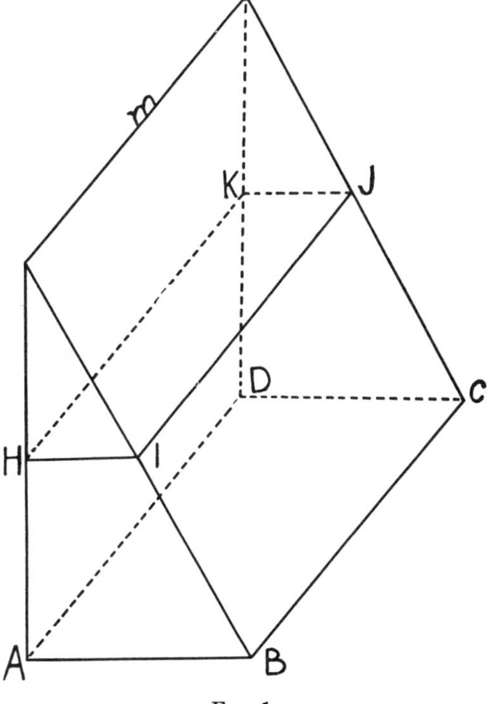

Fig. 1

A wedge may also be defined as a solid having a rectangular base $ABCD$; a line m parallel to a side of the rectangle and not in the plane of the rectangle; and its remaining faces determined by the totality of lines perpendicular to m and joining points of m to points of the rectangle. This definition may be readily modified to obtain the definition of a conoid. A *conoid* may be considered as a generalized wedge. Specifically, given any simple closed plane curve C and a line m not in the plane of C, but parallel to that plane, the totality of lines that are perpendicular to m and join points of m to the curve C together with the plane of C bound a conoid (Figure 2). In each of the above

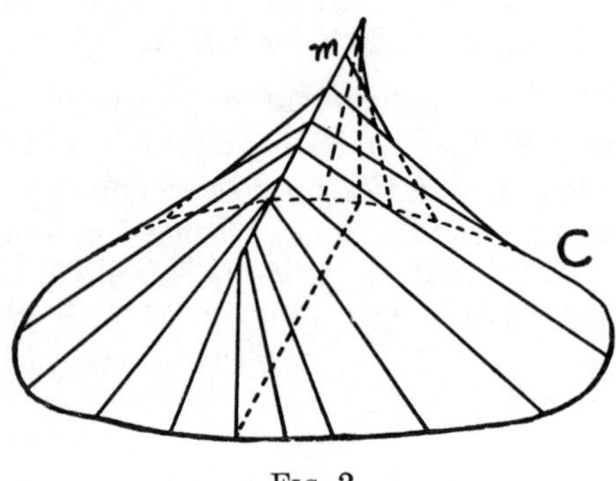

Fig. 2

types of solids, the area of the midsection may be taken as one-half the area of the first base and the second base may be taken as zero. Thus for all wedges and conoids

$$V = \frac{H}{6}\,(B_1 + 4B_1/2 + 0) = B_1 H/2 = BH/2.$$

IV. Volumes of a Sphere and a Spherical Segment

In the case of a sphere, we may take the first base at the south pole $(B_1 = 0)$, the second base at the north pole $(B_2 = 0)$, and the area of the midsection, i.e., the area enclosed by the equatorial circle as $M = \pi a^2$, and the height $H = 2a$. The prismoidal formula then gives

$$V = \frac{2a}{6}\,(0 + 4\pi a^2 + 0) = 4\pi a^3/3.$$

The prismoidal formula is also applicable to finding the volume of a spherical segment. It can be demonstrated that the radius r_m of the midsection is dependent upon the radius of the first base r_1, the radius of the second base r_2, and the altitude of the segment H. This dependence is expressed by the relationship

$$r_m{}^2 = r_1{}^2/2 + r_2{}^2/2 + H^2/4.$$

An application of the prismoidal formula now gives

$$V = \frac{H}{6}\left[\pi r_1{}^2 + 4\pi(r_1{}^2/2 + r_2{}^2/2 + H^2/4) + \pi r_2{}^2\right]$$

from which we may obtain the usual formula for the volume of a spherical segment

$$V = \frac{H}{6}\,\pi(3r_1{}^2 + 3r_2{}^2 + H^2).$$

V. Volumes of a Frustum of a Cone and a Frustum of a Pyramid

The prismoidal formula is also applicable to finding the volumes of a frustum of a cone and a frustum of a pyramid. It is not difficult to show that the area of the midsection (M) of a pyramid is related to the area (B_1) of the first base and (B_2) of the second base as indicated by the relationship

$$M = (\sqrt{B_1} + \sqrt{B_2})^2/4.$$

An application of the prismoidal formula now gives

$$V = \frac{H}{6}\left[B_1 + 4(\sqrt{B_1} + \sqrt{B_2})^2/4 + B_2\right]$$

from which we may obtain the usual formula for the volume of a frustum of a pyramid

$$V = \frac{H}{3}\,(B_1 + B_2 + \sqrt{B_1 B_2}).$$

In a similar manner the area of the midsection of a frustum of a cone is related to the areas and therefore the radii of the two bases as indicated by the relationship

$$M = \frac{\pi}{4}\,(r_1{}^2 + 2r_1r_2 + r_2{}^2).$$

In this case, the application of the prismoidal formula again results in the usual formula for the volume of a frustum of a circular cone

$$V = \frac{\pi H}{3}\,(r_1{}^2 + r_1r_2 + r_2{}^2).$$

VI. Volumes of Ellipsoids and Spheroids

The prismoidal formula is also applicable to the ellipsoids. Consider the ellipsoid

$$x^2/a^2 + y^2/b^2 + z^2/c^2 = 1.$$

As in the case of the sphere we may take $B_1 = B_2 = 0$ where the bases are in the planes $z = c$ and $z = -c$. The midsection of the ellipsoid then lies in the xy plane and is bounded by the ellipse $x^2/a^2 + y^2/b^2 = 1$, $z = 0$, which has area πab. Thus the prismoidal formula gives the usual result

$$V = \frac{2c}{6}\,(0 + 4\pi ab + 0) = 4\pi abc/3.$$

The volume of the oblate spheroid, the prolate spheroid, and the sphere can be considered as special cases of the volume of the ellipsoid for suitable values of a, b, and c. These three cases may be taken respectively as

$$V = 4\pi a^2 c/3 \quad \text{where} \quad a = b > c,$$
$$V = 4\pi a^2 c/3 \quad \text{where} \quad a = b < c, \quad \text{and}$$
$$V = 4\pi a^3/3 \quad \text{where} \quad a = b = c.$$

VII. Volume Common to Two Intersecting Right Circular Cylinders

Many calculus textbooks give the following problem: The axes of two right circular cylinders of radius a intersect at right angles. What is the volume common to the two cylinders? This problem can be easily solved by use of the prismoidal formula. Consider one-eighth of the total desired volume, for example, $ABCDE$ in

Figure 3 where ED and EC are radii of one cylinder, and ED and EA are radii of the other cylinder. Let us take the first

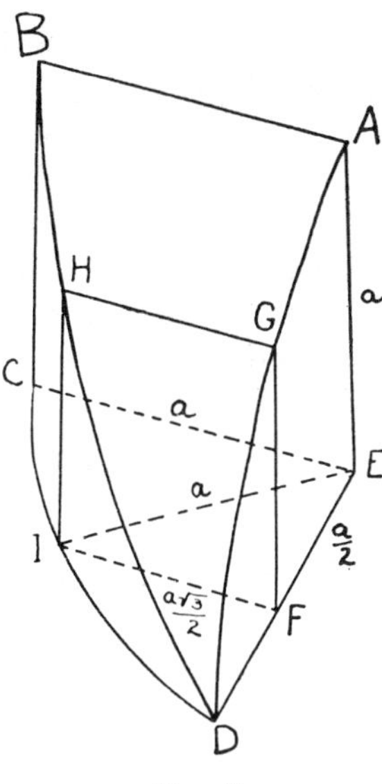

FIG. 3

base as the area of $ABCE$, the midsection as the area of $GHIF$, and the second base at D, $B_2 = 0$. We next use the fact that the faces EAD, ECD, and ECA are in mutually perpendicular planes, and the plane of $GHIF$ is parallel to that of $ABCE$. In the face ECD, we consider the right triangle EFI where $EI = a$, $EF = a/2$, and by the Pythagorean theorem $FI = a\sqrt{3}/2$. Similarly $GF = a\sqrt{3}/2$. It is then easy to see that $GHIF$ is a square of area $M = 3a^2/4$. Thus we have

$$V = \frac{a}{6}\,[a^2 + 4(3a^2/4) + 0] = 2a^3/3$$

as the volume of $ABCDE$. The desired volume is then

$$8(2a^3/3) = 16a^3/3.$$

The prismoidal formula may also be applied directly to the total volume common to two intersecting right circular cylinders [1]. In this case the cross-sectional areas are then squares with sides double those of the corresponding squares considered under the above method. Thus $H = 2a$, $B_1 = 0 = B_2$, $N = 4a^2$, and

$$V = \frac{2a}{6}(0 + 16a^2 + 0) = 16a^3/3.$$

VIII. Two Other Problems

Calculus textbooks often contain a problem similar to the following: In cutting down a tree 6 ft. in diameter, a cut is first made horizontally half way through the tree. A second cut is inclined at an angle of 45° to the horizontal and meets the first cut along a diameter of the tree. Compute the volume of the cylindrical wedge cut out (Figure 4).

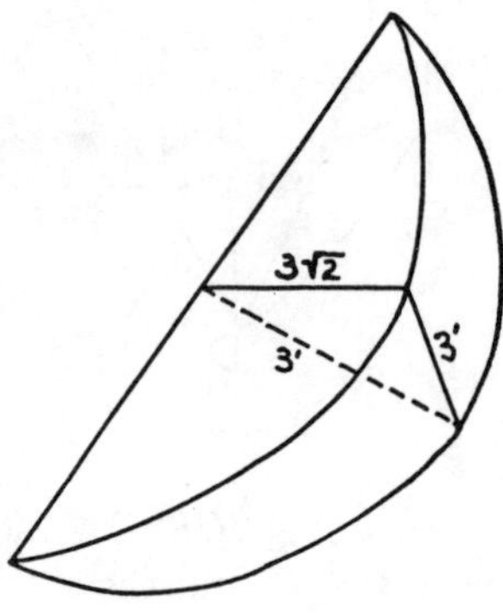

Fig. 4

The midsection of this solid is an isosceles right triangle with the legs 3 ft. Then $M = \frac{1}{2} \cdot 3 \cdot 3 = 9/2$ sq. ft., $B_1 = 0$, $B_2 = 0$, $H = 6$ ft., and the prismoidal formula gives

$$V = \frac{6}{6}[0 + 4(9/2) + 0] = 18 \text{ cu. ft.}$$

Another typical calculus problem involves the calculation of the volume in the first octant that is interior to the cylinder having the line $x = -z$, $y = 0$ as axis and intersecting the xy plane in the circle $x^2 + y^2 = a^2$ (Figure 5). This volume may be obtained using the prismoidal formula in which the first base is taken as the triangle in the xz plane, $B_1 = a^2/2$, the midsection is an isosceles right triangle, $M = 3a^2/8$; $B_2 = 0$; and $H = a$. Thus we have

$$V = \frac{a}{6}[a^2/2 + 4(3a^2/8) + 0] = a^3/3.$$

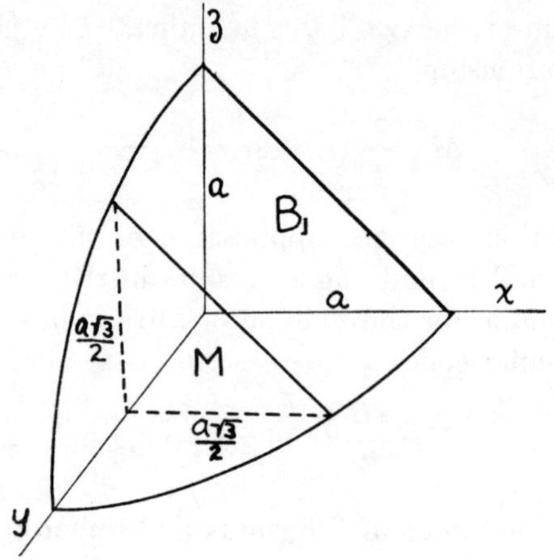

Fig. 5

Many other applications of the prismoidal formula could be given, however, we now turn to the task of determining whether or not the formula may be used for a given solid.

B. General Applicability of the Prismoidal Formula

In Section A several special cases of the application of the prismoidal formula were considered. The question still remains concerning the conditions for general application of the formula. Under what conditions can it be applied? Under what conditions can the formula not be applied? We now show that the prismoidal formula in general may be used for all solids having parallel bases and such that the area of every principal section is a polynomial of degree at most three in the distance of the section from a plane parallel to the bases. In other words, given a solid with parallel bases, we shall consider a coordinate system as in Figure 6 such that the area of any principal

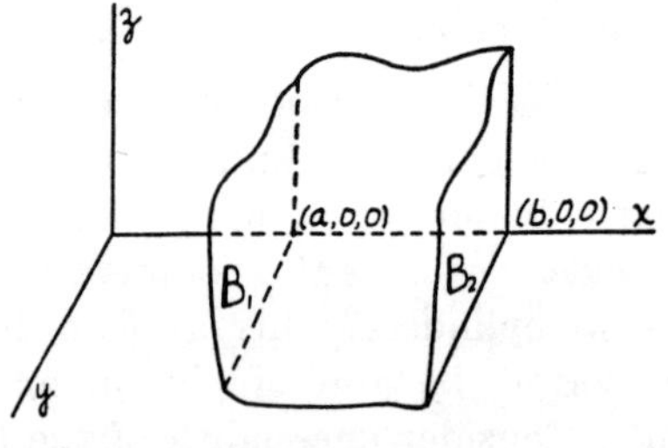

Fig. 6

section is a polynomial in the distance x of that section from a fixed plane parallel to it, and show that the prismoidal formula holds whenever $f(x)$ is a polynomial of degree not greater than three. The proofs are based upon the methods of the integral calculus.

Given a solid as described above, we have $B_1 = f(a)$, $B_2 = f(b)$, $M = f[(a+b)/2]$. Since $f(x)$ is a polynomial, it is integrable and the volume of the solid may be expressed in the form

$$V = \int_a^b f(x)dx.$$

Whenever $f(x)$ is a polynomial of degree at most three, i.e., $f(x) = Ax^3 + Bx^2 + Cx + D$, the above definite integral may be evaluated as follows to obtain the prismoidal formula.

$$V = \int_a^b (Ax^3 + Bx^2 + Cx + D)dx$$

$$= \frac{Ax^4}{4} + \frac{Bx^3}{3} + \frac{Cx^2}{2} + Dx \Big]_a^b$$

$$= \frac{A}{4}(b^4 - a^4) + \frac{B}{3}(b^3 - a^3)$$

$$+ \frac{C}{2}(b^2 - a^2) + D(b - a)$$

$$= (b-a)\left[\frac{A}{4}(b+a)(b^2+a^2)\right.$$

$$\left. + \frac{B}{3}(b^2 + ba + a^2) + \frac{C}{2}(b+a) + D\right]$$

$$= \frac{(b-a)}{6}\left[f(a) + f(b) + 4f\left(\frac{a+b}{2}\right)\right]$$

$$= \frac{(b-a)}{6}[B_1 + B_2 + 4M].$$

Thus the prismoidal formula holds whenever $f(x) = Ax^3 + Bx^2 + Cx + D$ for all x satisfying $a \leq x \leq b$. The coefficients A, B, C, D may be any real constants positive, negative, or zero.

If $f(x)$, $g(x)$, and $h(x)$ are any polynomials such that $f(x) = g(x) + h(x)$ then $\int f(x)dx = \int g(x)dx + \int h(x)dx$. Thus if $f_4(x) = a_0 + a_1x + a_2x^2 + a_3x^3 + a_4x^4$, we may consider $f_4(x) = f_3(x) + a_4x^4$. Since the prismoidal formula may be used whenever the areas of the principal sections may be represented by $f_3(x)$ it may be used for $f_4(x)$ if and only if it may be used for a_4x^4. In general, if the prismoidal formula may be used for $f(x)$, it may be used for $f(x) + cx^k$ if and only if it may be used for x^k.

If $f(x) = x^n$ we shall consider the cases $n = 0$ and $n > 0$. If $n = 0$ then $f(x)$ is 1, the volume is $b - a$, and the prismoidal formula applies. If $f(x) = x^n$ and $n > 0$, then by integration $V = (b^{n+1} - a^{n+1})/(n+1)$ whereas by the prismoidal formula we have

$$V = \frac{(b-a)}{6}\left[a^n + 4\left(\frac{b+a}{2}\right)^n + b^n\right].$$

These two values cannot be identically equal for all values of a and b unless the two coefficients of b^{n+1} are equal, i.e.,

$$\frac{1}{n+1} = \frac{1}{6}\left[1 + \frac{1}{2^{n-2}}\right] \quad \text{or} \quad 2^{n-2} = \frac{n+1}{5-n}.$$

Since this condition is satisfied only for $n = 1$, 2, or 3, the prismoidal formula does not hold for $n > 3$. In general, when $f(x)$ is a polynomial, the prismoidal formula holds if and only if $f(x)$ has the form $Ax^3 + Bx^2 + Cx + D$. This condition is satisfied for many common solids (Section A). It may fail to apply in general when $f(x)$ is a polynomial of degree greater than three or when $f(x)$ is not a polynomial, as in the case of a double cone. The above method of proof also applies whenever $f(x)$ has a Taylor series expansion. Thus the prismoidal formula does not in general apply if $f(x)$ is one of the functions having a Taylor series expansion that contains a term of degree greater than three. The trigonometric, logarithmic, and exponential functions are of this type.

It is a good exercise to verify that $f(x)$ has the proper form in each of the examples considered in Section A. For example, in the problem of two intersecting cylinders

(VII, Section A), all of the principal sections parallel to the base $ABCE$ are squares. If, in Figure 3, x is the distance of the principal section from the base $ABCE$, then one side of the square is $\sqrt{a^2-x^2}$. The area of a principal section is then (a^2-x^2) and is a polynomial in x of the form Ax^3+Bx^2+Cx+D where $A=C=0$, $B=-1$, and $D=a^2$. Thus the prismoidal formula is applicable to this particular solid.

The importance of these criteria for the use of the formula is illustrated by the following example. Given the parabola $y^2=4-x$, consider the volume bounded by the plane $x=0$ and the surface generated by revolving the parabola about the x-axis. We may take the base $B_1=4\pi$ in the plane $x=0$, $B_2=0$, and in general, $f(x)=\pi y^2=\pi(4-x)$. Thus $f(x)$ is a linear polynomial, and the prismoidal formula may be applied to give

$$V=\frac{4}{6}\left[4\pi+4(2\pi)+0\right]=8\pi.$$

Now consider the solid bounded by revolving the parabola $y^2=4-x$ about the y-axis. In this case we take the midsection in the plane $y=0$, and all principal sections are circles. The area of each principal section has the form

$$f(y)=\pi x^2=\pi(4-y^2)^2,$$

and the prismoidal formula does not apply since $f(y)$ is of the fourth degree.

A more complicated formula applicable when $f(y)$ is of degree at most five may be found in [3].

We have seen how the prismoidal formula is applicable to a large number of common solids, as well as to many that are not so common. Some explicit criteria for the application of the formula have been presented. Many students enjoy learning of broad generalizations that are applicable to a great number of special cases. Here then is an opportunity to teach for enjoyment as well as for practical application.

REFERENCES

1. Fehr, Howard F. *Secondary Mathematics*, Boston: D. C. Heath and Co., 1951, 327–29.
2. Frame, James S. *Solid Geometry*, New York: McGraw-Hill Book Co., Inc., 1948, 95–138.
3. ———. "Numerical Integration," *The American Mathematical Monthly*, L (April 1943), 244–50.
4. Swenson, John A. and Bakst, Aaron. *Notes on the Professionalized Matter in Senior High School Mathematics*. Part II, 22–25. New York, 1931.

An integraph

by George Kays, State Teachers College, Montclair, New Jersey

Finding an area enclosed by an irregular boundary is a problem that is quite often difficult or tedious when solved by the routine formal methods of the integral calculus. The device discussed here is a simplification of a type of instrument used to solve problems in integral calculus. Figure 1 illustrates such a problem, that of finding the area enclosed by the x-axis, the lines $x=a$, $x=b$, and the curve of the (unknown) function $y=F'(x)$.

If the function $F(x)$, an integral function of $F'(x)$, can be identified (or graphed), the measure of the required area would be $F(b)-F(a)$. $F(x)$ is, of course, any function which is continuous in the interval considered and has the slope of its graph numerically equal to the ordinate of $F'(x)$ for each value of x in that interval.

As an example consider the graph of the function shown in Figure 2. Note that the value of this $F'(x)$ is 1 for $x=0$, and thus the slope of $F(x)$ is 1 for $x=0$. The curve of $F(x)$ can then be approximated near $x=0$ by any one of the lines with slope 1 drawn through the points P, Q, R, S. Similarly, the value of $F'(1)$ is 2,

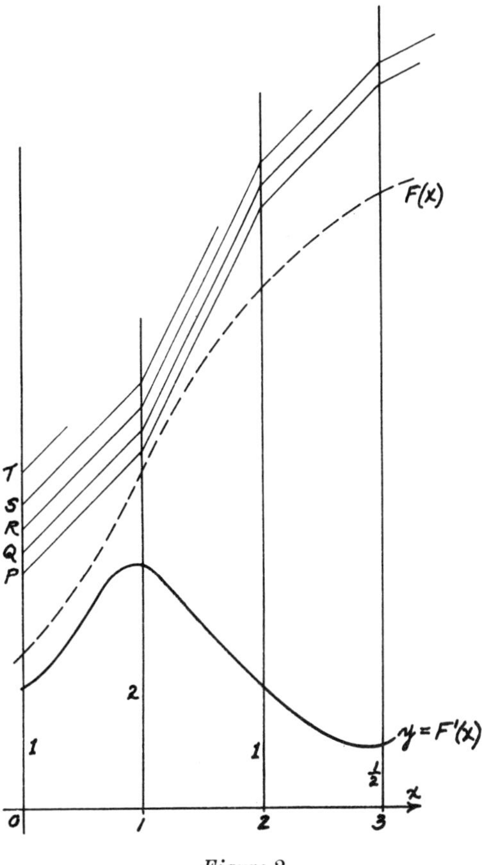

Figure 2

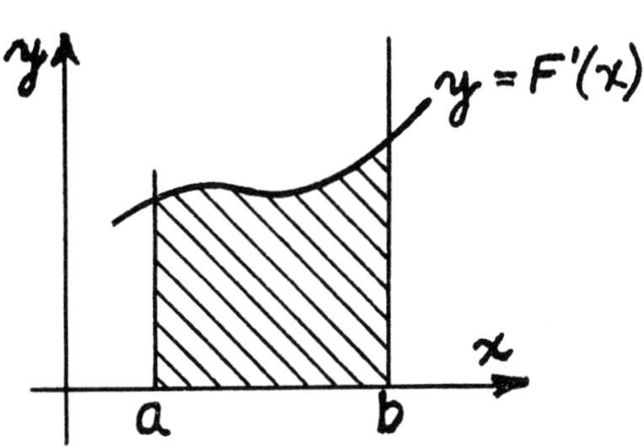

Figure 1

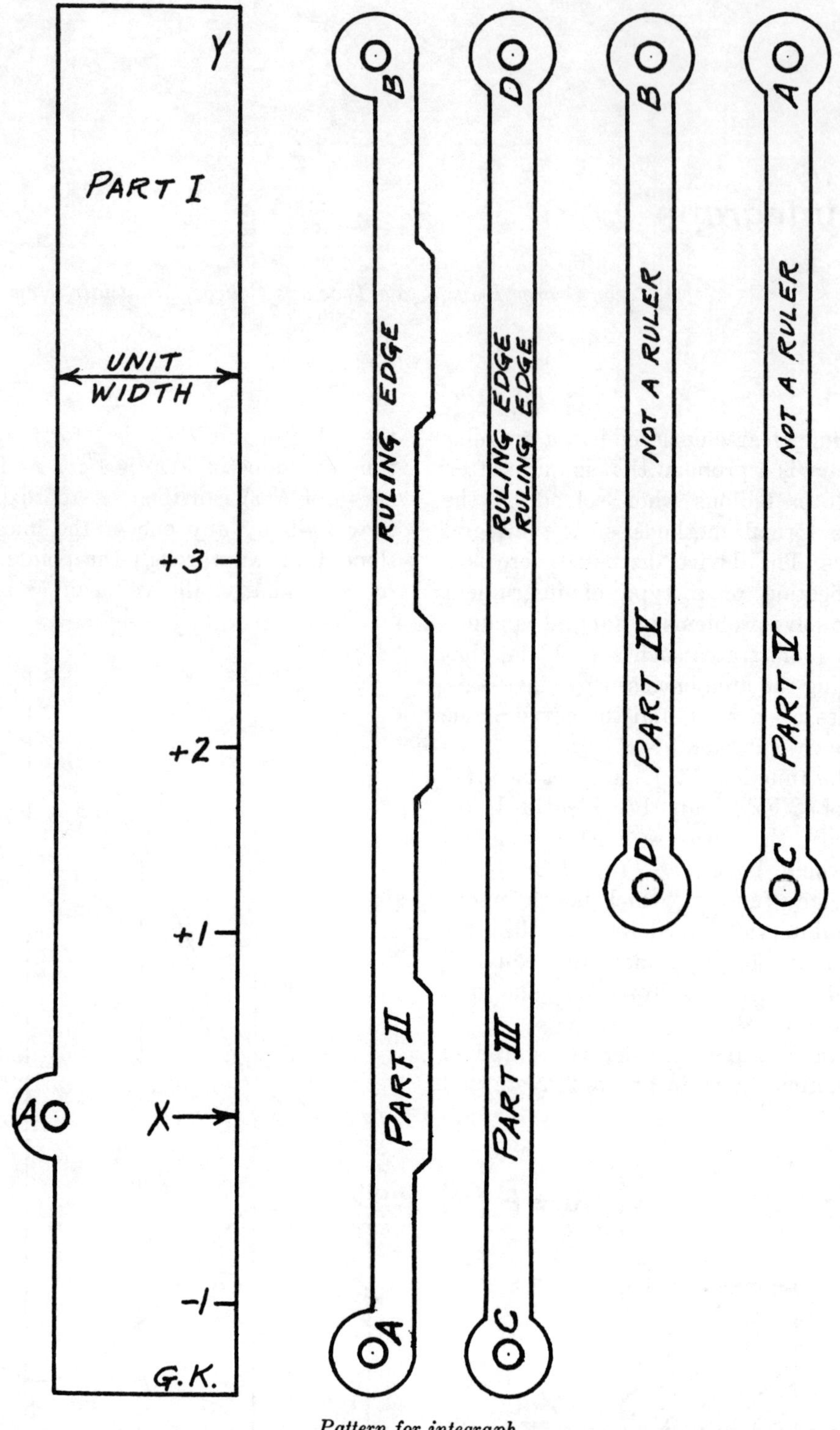

Pattern for integraph

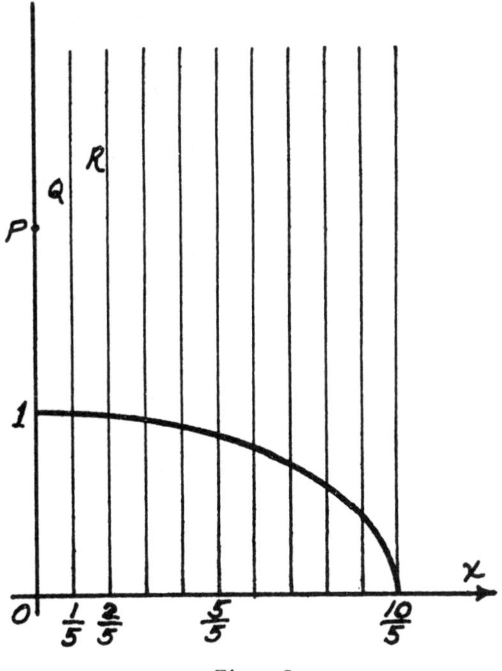

Figure 3

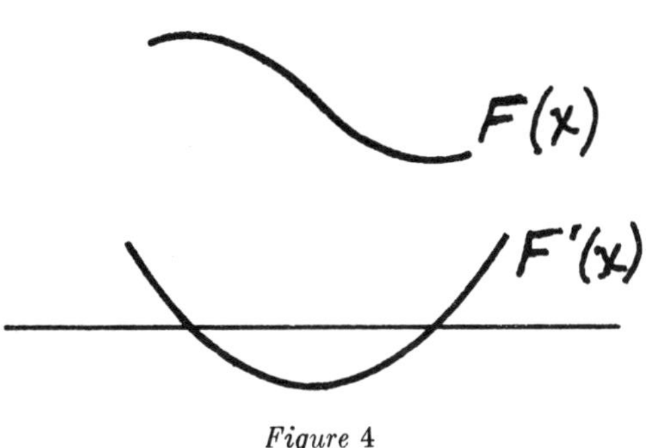

Figure 4

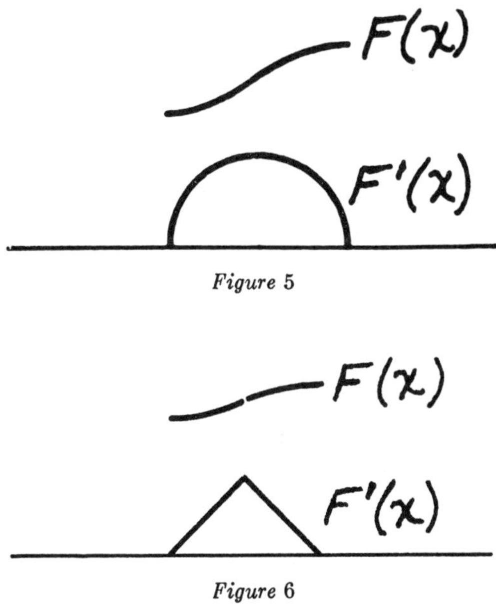

Figure 5

Figure 6

and the curve of $F(x)$ can be approximated by any of the lines with slope 2. Continuing, the graph is further approximated by the lines with slope 1 and slope $\frac{1}{2}$ in the vicinity of $x=2$ and $x=3$ respectively.

If one of the functions $F(x)$ were actually known, as shown by the dotted curve, then the area beneath $F'(x)$ from $x=0$ to $x=3$ could be found by $F(3)-F(0)$. Since this is not known, the broken line through Q (or P, or R, etc.) may be used in place of the curve. Using intervals of $\frac{1}{2}$ unit, $\frac{1}{4}$ unit, etc., will increase the accuracy of the approximation to desired precision.

At this point the reader is urged to construct the device described at the end of this article.

In Figure 3 a curve $F'(x)$ is shown. The curve is an ellipse, and the area of the quadrant is $\frac{1}{2}\pi$, or about 1.57 square inches. To approximate this area place the integraph so that AX aligns with the x-axis, XY with the y-axis, and the ruling edge of AB passes through the intersection of $F'(x)$ and XY. Using the parallel ruler CD construct a line, say PQ, which will approximate $F(x)$ for the first interval. Next place the integraph so that the line AX remains on the x-axis, XY is aligned with the line $x=\frac{1}{5}$, AB passes through the intersection of $F'(x)$ and the line $x=\frac{1}{5}$. From Q the approximation to $F(x)$ is extended through the second interval to R by means of the parallel ruler CD. Continue the construction of $PQR\ldots$ until a sequence of connected straight line segments has been drawn to represent the curve of $F(x)$. To find the area beneath $F'(x)$ find the value of $F(2)-F(0)$.

Other problems that might be of interest to the reader are illustrated in Figures 4, 5, 6.

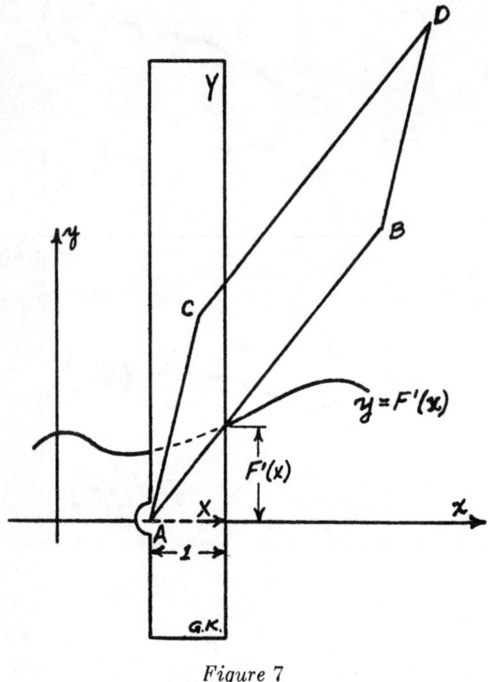

Figure 7

DIRECTIONS FOR ASSEMBLING THE
INTEGRAPH

Trace the pattern, including labels, transferring it to light cardboard or O'Tag. Punch the holes, marked O. Use an eyelet set and eyelets to assemble the parts. Each joint is made by placing all of the same letters at one joint, the higher numbered parts being above the lower. For instance, joint A is made by fastening the ends of parts I, II, and V which are marked with an A. Part II is above Part I, Part V is above both of the others. The eyelets should be carefully set since, if they are set too tightly, there will be no motion.

Note: For some problems an additional parallel ruler may be added above CD for greater mobility and flexibility.

Proof: If the integraph is placed on the graph as directed in the discussion, then it is obvious that the pointer AB (and therefore CD) has slope $F'(x)/1$, or $F'(x)$ (see Fig. 7).

Bibliography: *Numerical and Mechanical Methods*

Harkin, J. B. The Limit Concept on the Geoboard. 65 (January 1972): 13–17.
Remarks on finding areas of simple closed curves using a numerical-graphical technique based on Pick's formula.

9

Infinite Series

Standard approaches to infinite series frequently leave students perplexed and confused. Their intuition fails them. They cannot really see why certain series are said to converge and others, very similar in appearance, diverge. They fail to see how and where infinite series can be of use to any but the most theoretical mathematician. The papers presented here and also listed in the bibliography illustrate some ways of handling the difficulties inherent in teaching this topic.

Wollan provides insight into the historical development of the theory of infinite series. Then Pavlovich shows how a computer can be used as a guide in resolving convergence questions. He programs a computer to provide clues in instances where standard convergence tests either fail or are difficult to apply.

These articles represent some of the ways for making the theory of infinite series more meaningful to the student. They can serve as a stimulus for further innovative exploration on the part of both students and faculty.

MACLAURIN and TAYLOR
and Their Series

By G. N. Wollan, Purdue University, Lafayette, Indiana

BROOK Taylor (1685–1731) was an eminent British mathematician whose relatively short lifetime roughly coincided with the latter half of the lifetime of Isaac Newton. He was an unusually talented youth with a wide range of interests, including music and art as well as mathematics and philosophy. A derivation of the series which bears his name was included in his principal mathematical work, a book *Methodus Incrementorum directa et inversa* which was published in London in 1715. At that time the mathematical world was deeply embroiled in the Newton-Leibniz priority controversy. Taylor was an ardent supporter of Newton's claim, and the book is filled with biased references to Newton and his work, no mention being made of other contributors. The book was primarily devoted to the development of a branch of mathematics known today as the calculus of finite differences. Much of its content was then new, and the book acquired considerable renown in spite of the fact that the author's obscure style, his complicated notation, and other defects made it very difficult to read. Put into modern notation, Taylor's derivation of his series as given in *Methodus Incrementorum* was approximately as follows.

SUPPOSE $y(x)$ is a function of x, Δx is an arbitrary real number, and

$$\Delta y(x) = y(x + \Delta x) - y(x)$$

is the change in the value of the function

corresponding to a change of Δx in the value of the argument x. We also define higher order differences of the function:

$$\begin{aligned}
\Delta^2 y(x) &= \Delta(\Delta y(x)) \\
&= \Delta y(x + \Delta x) - \Delta y(x), \\
\Delta^3 y(x) &= \Delta(\Delta^2 y(x)) \\
&= \Delta^2 y(x + \Delta x) - \Delta^2 y(x), \\
\Delta^n y(x) &= \Delta(\Delta^{n-1} y(x)) \\
&= \Delta^{n-1} y(x + \Delta x) - \Delta^{n-1} y(x).
\end{aligned}$$

It follows of course that if $z(x)$ is any other function of x then

$$\Delta(y(x) + z(x)) = \Delta y(x) + \Delta z(x).$$

Now then, a being an arbitrary number, we have

$$\begin{aligned}
y(a + \Delta x) &= y(a) + \Delta y(a); \\
y(a + 2\Delta x) &= y(a + \Delta x) + \Delta y(a + \Delta x) \\
&= (y(a) + \Delta y(a)) + \Delta(y(a) + \Delta y(a)) \\
&= y(a) + \Delta y(a) + \Delta y(a) + \Delta^2 y(a) \\
&= y(a) + 2\Delta y(a) + \Delta^2 y(a); \\
y(a + 3\Delta x) &= y(a + 2\Delta x) + \Delta y(a + 2\Delta x) \\
&= (y(a) + 2\Delta y(a) + \Delta^2 y(a)) \\
&\qquad + \Delta(y(a) + 2\Delta y(a) + \Delta^2 y(a)) \\
&= y(a) + 2\Delta y(a) + \Delta^2 y(a) + \Delta y(a) \\
&\qquad + 2\Delta^2 y(a) + \Delta^3 y(a) \\
&= y(a) + 3\Delta y(a) + 3\Delta^2 y(a) + \Delta^3 y(a);
\end{aligned}$$

and, in general,

$$\begin{aligned}
y(a + n\Delta x) &= y(a + (n-1)\Delta x + \Delta y(a + (n-1)\Delta x) \\
&= y(a) + n\Delta y(a) + \frac{n(n-1)}{2!}\Delta^2 y(a) \\
&\qquad + \frac{n(n-1)(n-2)}{3!}\Delta^3 y(a) \\
&\qquad + \ldots + \Delta^n y(a).
\end{aligned}$$

Letting $n\Delta x = v$ then gives

$$y(a + v) = y(a) + \frac{v}{\Delta x}\, y(a)$$

$$+ \frac{\dfrac{v}{\Delta x}\left(\dfrac{v}{\Delta x} - 1\right)}{2!}\, \Delta^2 y(a)$$

$$+ \frac{\dfrac{v}{\Delta x}\left(\dfrac{v}{\Delta x} - 1\right)\left(\dfrac{v}{\Delta x} - 2\right)}{3!}\, \Delta^3 y(a)$$

$$+ \ldots + \Delta^n y(a)$$

$$= y(a) + v\,\frac{y(a)}{\Delta x} + \frac{v(v - \Delta x)}{2!}\,\frac{\Delta^2 y(a)}{(\Delta x)^2}$$

$$+ \frac{v(v - \Delta x)(v - 2\Delta x)}{3!}\,\frac{\Delta^3 y(a)}{(\Delta x)^3} + \ldots + \frac{\Delta^n y(a)}{(\Delta x)^n}.$$

At this point Taylor argued that if we consider v fixed and let n become infinite, then Δx approaches 0,

$$\frac{\Delta^K y(a)}{(\Delta x)^K}$$

and

$$\frac{v(v - \Delta x)(v - 2\Delta x) \ldots (v - (K - 1)\Delta x)}{K!}$$

become

$$\frac{d^K y(a)}{dx^K}$$

and

$$\frac{v^K}{K!},$$

respectively, for each K, and the number of terms becomes infinite so that we get

$$y(a + v) = y(a) + v\,\frac{dy(a)}{dx} + \frac{v^2}{2!}\,\frac{d^2 y(a)}{dx^2}$$

$$+ \frac{v^3}{3!}\,\frac{d^3 y(a)}{dx^3} + \ldots,$$

which, of course, is Taylor's series.

It was this sort of fuzzy argument regarding limits which left not only Taylor but also Newton, Leibniz, and all their contemporaries vulnerable to the attack of Bishop Berkeley and led later mathematicians to develop the proofs that students learn today.

COLIN Maclaurin (1698–1746), a brilliant Scottish mathematician who became pro-

fessor of mathematics at the University of Aberdeen at age nineteen on the basis of a competitive examination, devised a different derivation of the series which appeared in his book *Treatise on Fluxions*, published in Edinburgh in 1742. Maclaurin's argument was intended to meet the objections raised by Berkeley and in modern notation was essentially as follows.

Assume that

$$y(x) = A_0 + A_1 x + A_2 x^2 + A_3 x^3 + \ldots,$$

where the coefficients A_0, A_1, A_2, A_3, $\ldots$ are fixed numbers whose values are to be determined. Then

$$\frac{dy(x)}{dx} = A_1 + 2A_2 x + 3A_3 x^2 + \ldots,$$

which, setting $x = 0$, gives

$$\frac{dy(0)}{dx} = A_1.$$

Similarly,

$$\frac{d^2 y(x)}{dx^2} = 2A_2 + 3\cdot 2A_3 + 4\cdot 3A_4 x^2 + \ldots,$$

so that

$$\frac{d^2 y(0)}{dx^2} = 2A_2,$$

and it is clear that

$$\frac{d^n y(x)}{dx^n} = n!A_n + (n + 1)n(n - 1) \ldots 2A_{n+1} x$$

$$+ (n + 2)(n + 1) \ldots 3A_{n+2} x^2 + \ldots,$$

from which

$$\frac{d^n y(0)}{dx^n} = n!A_n.$$

Thus

$$y(x) = y(0) + x\,\frac{dy(0)}{dx} + \frac{x^2}{2!}\,\frac{d^2 y(0)}{dx^2}$$

$$+ \frac{x^3}{3!}\,\frac{d^3 y(0)}{dx^3} + \ldots.$$

It is apparent that in this argument the following questions are ignored: What functions have such series representations? For what values of x does the series represent the function? Does term-by-term differentiation of the series yield the derivative of the given function?

Neither Taylor nor Maclaurin has a valid claim to having discovered the series to which their names are attached since they appeared in a previously published work of John Bernoulli with which both men were probably familiar. However, both men made significant contributions to mathematics and probably deserve the measure of immortality which the association of their names with the series has given them.

SOURCE

M. Cantor, *Geschichte der Mathematik*, Vol. III.

INDUCTIVE THINKING USING THE COMPUTER

By **JOSEPH P. PAVLOVICH**

Shady Side Academy
Pittsburgh, Pennsylvania

IN ONE of my recent classes in the calculus course, we had the following problem (Crowell and Slesnick 1968):

> 3. Classify each of the following series as absolutely convergent, conditionally convergent, or divergent. Show how you obtain your answer.
>
> .
> .
> .
>
> d. $\qquad \sum_{n=1}^{\infty} \dfrac{\ln n}{n^2}$

When we came upon this problem, we had been working with infinite series for over two weeks, and at this point we had a number of theorems at our disposal to investigate infinite series. Included in this list of theorems and definitions from Crowell and Slesnick were the following:

1. *An infinite series $\sum_{i=m}^{\infty} a_i$ is said to be absolutely convergent if the corresponding series of absolute values $\sum_{i=m}^{\infty} |a_i|$ is convergent.*

2. *If a series $\sum_{i=m}^{\infty} a_i$ converges but $\sum_{i=m}^{\infty} |a_i|$ does not, then we say that $\sum_{i=m}^{\infty} a_i$ is conditionally convergent.*

3. *Every nonnegative series $\sum_{i=m}^{\infty} a_i$ either converges or satisfies $\sum_{i=m}^{\infty} a_i = \infty$.*

4. *If $\sum_{i=m}^{\infty} a_i$ is a nonnegative series, then the corresponding sequence $\{s_n\}$ of partial sums is an increasing sequence.*

5. *If $\sum_{i=m}^{\infty} a_i$ converges, then $\lim_{n \to \infty} a_n = 0$.*

6. *If $\{s_n\}$ is an increasing sequence of real numbers, then either it is bounded above and therefore converges or else $\lim_{n \to \infty} s_n = \infty$.*

7. *Integral Test. Let f be a function that is nonnegative and decreasing on the interval $[m, \infty)$. Then the infinite series $\sum_{i=m}^{\infty} a_i$, defined by $a_i = f(i)$, for every $i \geq m$, is convergent if and only if the improper integral $\int_m^{\infty} f(x)dx$ is convergent.*

8. *Comparison Test. If $\sum_{i=m}^{\infty} a_i$ is a nonnegative series and if $\sum_{i=m}^{\infty} b_i$ is a convergent series with $a_i \leq b_i$ for every $i \geq m$, then $\sum_{i=m}^{\infty} a_i$ converges and $\sum_{i=m}^{\infty} a_i \leq \sum_{i=m}^{\infty} b_i$.*

9. *If the infinite series $\sum_{i=m}^{\infty} a_i$ is absolutely convergent, then it is convergent.*

10. *Ratio Test. Let $\sum_{i=m}^{\infty} a_i$ be an infinite series for which*

$$\lim_{n \to \infty} \frac{|a_{n+1}|}{|a_n|} = q \ (\text{or } \infty).$$

(i) *If $q < 1$, then the series is absolutely convergent.*

(ii) *If $q > 1$ (including $q = \infty$), then the series is divergent.*

(iii) *If $q = 1$, then the series may either converge or diverge; i.e., the test fails.*[1]

As our first approach to the problem, we considered the Ratio Test, as follows:

First,

$$\lim_{n \to \infty} \frac{\left| \dfrac{\ln (n+1)}{(n+1)^2} \right|}{\left| \dfrac{\ln (n)}{n^2} \right|} = \lim_{n \to \infty} \frac{\dfrac{\ln (n+1)}{(n+1)^2}}{\dfrac{\ln (n)}{n^2}}$$

for $n \geq 1$. Now,

$$\lim_{n \to \infty} \frac{\dfrac{\ln (n+1)}{(n+1)^2}}{\dfrac{\ln (n)}{n^2}}$$

$$= \lim_{n \to \infty} \left[\frac{\ln (n+1)}{(n+1)^2} \cdot \frac{n^2}{\ln (n)} \right]$$

$$= \lim_{n \to \infty} \left[\frac{n^2}{(n+1)^2} \cdot \frac{\ln (n+1)}{\ln (n)} \right]$$

$$= \lim_{n \to \infty} \frac{n^2}{(n+1)^2} \cdot \lim_{n \to \infty} \frac{\ln (n+1)}{\ln (n)}.$$

By L'Hospital's rule, we have:

$$\lim_{n \to \infty} \frac{n^2}{(n+1)^2} = \lim_{n \to \infty} \frac{2n}{2n+2} = 1$$

and

---

1. Reprinted from *Calculus with Analytic Geometry* by R. H. Crowell and W. E. Slesnick, © 1968, by permission of the publisher, W. W. Norton and Co., New York, New York.

$$\lim_{n\to\infty} \frac{\ln(n+1)}{\ln(n)} = \lim_{n\to\infty} \frac{\frac{1}{n+1}}{\frac{1}{n}}$$

$$= \lim_{n\to\infty} \frac{n}{n+1} = 1.$$

Therefore,

$$\lim_{n\to\infty} \frac{n^2}{(n+1)^2} \cdot \lim_{n\to\infty} \frac{\ln(n+1)}{\ln(n)} = 1\cdot 1 = 1,$$

and the Ratio Test as given in theorem 10 above fails!

We then tried the Comparison Test in conjunction with the integral test and thus searched for a series $\sum_{i=m}^{\infty} b_i$ such that $\frac{\ln(n)}{n^2} \le b_i$ for $i \ge m$ and also such that $\lim_{c\to\infty} \int_1^c f(b_x)dx$ exists. An obvious series to consider was $\sum_{n=1}^{\infty} \frac{\ln(n)}{n}$, since $\frac{\ln(n)}{n^2} \le \frac{\ln(n)}{n}$ for $n \ge 1$.

But,

$$\lim_{c\to\infty} \int_1^c \frac{\ln(x)}{x}\, dx$$

$$= \lim_{c\to\infty} \left[\frac{1}{2}(\ln(x))^2 \right]\Big|_1^c$$

$$= \lim_{c\to\infty} \frac{1}{2}(\ln(c))^2 - \lim_{c\to\infty} \frac{1}{2}(\ln(1))^2$$

$$= \infty.$$

However, knowing that the series $\sum_{n=1}^{\infty} \frac{\ln(n)}{n}$ diverges told us nothing about the convergence or divergence of

$$\sum_{n=1}^{\infty} \frac{\ln(n)}{n^2}.$$

We then realized that

$$\lim_{x\to\infty} \frac{\ln(x)}{x^2} = 0,$$

since by L'Hospital's rule we have:

$$\lim_{x\to\infty} \frac{\ln(x)}{x^2} = \lim_{x\to\infty} \frac{\frac{1}{x}}{2x} = \lim_{x\to\infty} \frac{1}{2x^2} = 0.$$

However, the converse of theorem 5 above

is false. (Recall that the harmonic series $\sum_{n=1}^{\infty} \frac{1}{n}$ diverges, even though $\lim_{n\to\infty} \frac{1}{n} = 0$.) Thus, knowing that

$$\lim_{x\to\infty} \frac{\ln(x)}{x^2} = 0$$

is true could not help us decide on the convergence or divergence of $\sum_{n=1}^{\infty} \frac{\ln(n)}{n^2}$.

We then turned to the computer to see if it could shed some light on the convergence or divergence of $\sum_{n=1}^{\infty} \frac{\ln(n)}{n^2}$. Of course, we knew that the machine would not give us a definite or complete answer (the computer can handle large numbers, but it cannot go out to infinity), but it could possibly give us a hint about the solution to the problem.

We wrote a short program in BASIC and then received the output shown in figure 1 from a time-sharing computer system ASR 33 terminal.

Does theorem 6 above apply? After 10,000 partial sums of the infinite series $\sum_{n=1}^{\infty} \frac{\ln(n)}{n^2}$, could we be convinced that there exists an upper bound to the series? We had not proved anything, but we were inclined to believe that $\sum_{n=1}^{\infty} \frac{\ln(n)}{n^2}$ converges.

We then tried a different approach to the problem, using the integral test. We had on file in the computer a program (Pavlovich and Tahan 1971) that would enable us to find the approximate area under the curve of any continuous function by four different methods: the rectangular, the trapezoidal, the midpoint, and Simpson's rules.

We then used that program for the function $\frac{\ln(x)}{x^2}$ and considered the successive closed intervals [1,100], [1,200], [1,300], [1,400], [1,500], [1,1000], and [1,2000], with appropriate subintervals in each case such that $|x_{i+1} - x_i| = 0.1$.

The computer output shown in figure 2

```
1Ø FOR N=1 TO 1ØØØØ
2Ø     LET A=LOG(N)/N↑2
3Ø     LET B=B+A
4Ø     IF N/1ØØ−INT(N/1ØØ)=Ø THEN 7Ø
5Ø     IF N=1ØØØØ THEN 9Ø
6Ø NEXT N
7Ø PRINT B;
8Ø GOTO 5Ø
9Ø END
RUN
```

.881726	.9Ø6123	.915234	.92ØØ88	.923131	.925229	.926768
.927948	.928883	.929644	.93Ø276	.93Ø8Ø9	.931266	.931661
.932ØØ8	.932314	.932586	.932830	.933Ø5Ø	.933249	.93343Ø
.933596	.933749	.933889	.934Ø19	.93414Ø	.934252	.934357
.934455	.934547	.934633	.934714	.934791	.934863	.934931
.934996	.935Ø58	.935116	.935172	.935225	.935276	.935324
.93537Ø	.935415	.935457	.935498	.935537	.935574	.93561Ø
.935645	.935678	.935711	.935742	.935772	.935801	.935829
.935856	.935882	.9359Ø7	.935932	.935956	.935979	.936ØØ1
.936Ø23	.936Ø44	.936Ø64	.936Ø84	.9361Ø4	.936122	.936141
.936159	.936176	.936193	.9362Ø9	.936225	.936241	.936256
.936271	.936286	.936300	.936314	.936327	.936341	.936354
.936366	.936379	.936391	.936403	.936414	.936426	.936437
.936448	.936458	.936469	.936479	.936489	.936499	.936508
.936518	.936527					

Fig. 1

suggests that the successive approximations of the area under the curve of $\frac{\ln (x)}{x^2}$ from $x = 1$ to $x = B$ for $B = 100, 200, 300, 400, 500, 1000, 2000$ converges to the number 1. Hence, this second approach convinced us a bit more that $\sum_{n=1}^{\infty} \frac{\ln (n)}{n^2}$ is a convergent series.

Of course, these two computer approaches to the problem at hand do not prove convergence of the series $\sum_{n=1}^{\infty} \frac{\ln (n)}{n^2}$, but the computer output offers strong evidence to suggest that this is the case. Also, the power of the computer was certainly exhibited as it typed out both the successive partial sums of the series for $n = 100, 200, 300, \cdots , 10000$, and the successive values of the area under the graph of the function $\frac{\ln (x)}{x^2}$ for $A = 1$ and $B = 100, 200, 300, 400, 500, 1000,$ and 2000.

We then considered the integral test as given in theorem 7 above, where $\sum_{n=1}^{\infty} \frac{\ln (n)}{n^2}$ can be compared with the improper integral

$$\int_1^{\infty} \frac{\ln (x)}{x^2} \, dx = \lim_{c \to \infty} \int_1^c \frac{\ln (x)}{x^2} \, dx,$$

and then tried to find the value of this integral using traditional methods.

Using the method of integration by parts for $\lim_{c \to \infty} \int_1^c \frac{\ln (x)}{x^2} \, dx$, we let $u = \ln (x)$ and $dv = x^{-2} \, dx$. Then $du = \frac{1}{x} \, dx$ and $v = -x^{-1}$. Thus

$$\lim_{c \to \infty} \int_1^c \frac{\ln (x)}{x^2} \, dx$$

OUTPUT

NUMBER OF SUBINTERVALS	RECTANGULAR RULE	TRAPEZOIDAL RULE	MIDPOINT RULE	SIMPSON'S RULE
1000	.943112	.943135	.944354	.943935
2000	.967680	.967687	.968918	.968495
3000	.976827	.976830	.978065	.977640
4000	.981694	.981696	.982933	.982508
5000	.984743	.984744	.985983	.985557
10000	.991264	.991264	.992505	.992079
20000	.994871	.994871	.996113	.995686

Fig. 2

$$= \lim_{c \to \infty} \left[-\frac{\ln (x)}{x} + \int_1^c \frac{dx}{x^2} \right]$$

$$= \lim_{c \to \infty} \left[-\frac{\ln (x)}{x} - \frac{1}{x} \right]\Big|_1^c$$

$$= (-0 - 0) - (-0 - 1)$$

$$= 1.$$

Since

$$\lim_{c \to \infty} \int_1^c \frac{\ln (x)}{x^2} \, dx = 1,$$

we have indeed shown that the series $\sum_{n=1}^{\infty} \frac{\ln (n)}{n^2}$ converges.

Investigating the series $\sum_{n=1}^{\infty} \frac{\ln (n)}{n^2}$ by computer methods not only gave my students additional insight into infinite series but also gave them an appreciation for another powerful method by which they can investigate some problems in calculus—by what Professor G. Polya (1957) calls "shrewd guesswork" and intuition based on inductive methods.

Perhaps the computer can offer never-before-available inductive methods to help solve problems whose solutions might not be readily apparent by deductive methods. This paper attempted to show one such problem. Time and experience may show more.

BIBLIOGRAPHY

Crowell, R. H., and W. E. Slesnick. *Calculus with Analytic Geometry*. New York: W. W. Norton & Co., 1968.

Pavlovich, J. P., and T. E. Tahan. *Computer Programming in BASIC*. San Francisco: Holden-Day, 1971.

Polya, G. *How to Solve It*. 2d ed. Garden City, N.Y.: Doubleday & Co., 1957.

Bibliography: *Infinite Series*

Sanford, Vera. Brook Taylor, 1685–1731. 27 (January 1934): 60–61.

______. Colin Maclaurin, 1698–1746. 27 (March 1934): 155–56.

Berger, Emil J. A Model for Teaching Infinite Series to High School Students. 47 (February 1954): 101–5.

Remarks concerning the construction and operation of a physical model that can be used to illustrate series summation, convergence, and divergence.

Tehan, Vincent, S.C.N. The Use of the Taylor Series in Change of Bases. 59 (February 1966): 134–37.

Remarks demonstrating the use of the Taylor series theory in the conversion from one base of number representation to another.

Rasof, Bernard. Error Analysis without Calculus. 61 (January 1968): 2–11.

Remarks illustrating the application of infinite series to problems in error analysis.

Young, Worth J. The Bouncing Ball Does Come to Rest. 63 (May 1970): 391–92.

Remarks concerning the question of how a bouncing ball can make an infinite number of bounces in a finite length of time.

10

Special Numbers: e and π

The numbers e and π occur with great frequency in calculus and more advanced mathematics. The average calculus text, although presenting necessary facts and definitions, fails to provide insight into the significance and pervasiveness of these numbers throughout mathematics. The material in this section serves to bridge this gap so evident in current calculus texts.

The article by Baravalle is an expository paper describing the properties of e, its role in differential equations, and its significance in the study of trigonometric, logarithmic, and hyperbolic functions. The second article, extensively researched by Wrench, deals with attempts through the ages to compute decimal approximations to π. The role of π in computing areas bounded by curves, in calculating volumes of solids, and in the formulas of probability, statistics, and physics is described in a comprehensive article by Baravalle. Some expansions for π in terms of infinite series as well as continued fractions are also given. Finally, Schaumberger gives a detailed discussion of the definition of e.

This section can aid the instructor in illustrating interconnections among various mathematical disciplines. It can also be used as a starting point for the further examination of these numbers.

The Number e—The Base of Natural Logarithms

By H. V. Baravalle
Adelphi College, Garden City, N. Y.

Among the outstanding constants in mathematics e ranks first, even before π, in regard to both the variety of its mathematical implications and the amount of its practical applications. The most outstanding qualities of e show up in the calculus. e is the basis of the exponential function which is unchanged through differentiation. For $y = e^x$ we get $y' = e^x$; $y'' = e^x$; $\cdots y^n = e^x$. All derivatives of $y = e^x$ are equal to the original function. The general solution of the differential equation $\dfrac{dy}{dx} = y$ is $y = C \cdot e^x$ (separating the variables we get $\dfrac{dy}{y} = dx$; ln $y = x + c$, $y = e^{x+c}$ and for $e^c \equiv C$: $y = C \cdot e^x$). Therefore, all possible functions which are equal to their first derivative and therefore also to their higher derivatives are products of the exponential function with the basis e and a constant factor,

An immediate consequence of this fact regarding e is the expression of e obtained through expansion of the function $y = e^x$ according to MacLaurin's series:

$$f(x) = f(0) + \frac{f'(0)}{1!} x + \frac{f''(0)}{2!} x^2 + \frac{f'''(0)}{3!} x^3 + \cdots \frac{f^{(n)}(0)}{n!} x^n + \cdots .$$

As in the case of $f(x) = e^x$ the derivatives are $f'(x) = f''(x) = f^{(n)}(x) = e^x$ and we obtain for $f(0) = e^0 = 1$; for $f'(0) = f''(0) = f^{(n)}(0) = 1$. For e^x we get the power series

$$e^x = 1 + \frac{x}{1!} + \frac{x^2}{2!} + \frac{x^3}{3!} + \cdots \frac{x^n}{n!} + \cdots$$

substituting for x the value 1 we have

$$e = 1 + \frac{1}{1!} + \frac{1}{2!} + \frac{1}{3!} + \frac{1}{4!} + \cdots \frac{1}{n!} \cdots .$$

which being a convergent series leads to the definite value of $e = 2.718281828459 \cdots$.

A practical application of the described quality of e is Euler's method to solve linear differential equations, both of the second and of higher orders, through the substitution $y = e^{rx}$. As the linear differential equations of the second order include the outstanding differential equation for harmonic vibrations: $\dfrac{d^2y}{dx^2} = -a^2y$ there is an almost unlimited field of applications in mechanics, in acoustics, electricity, etc. With the substitution $y = e^{rx}$ we obtain $\dfrac{d^2y}{dx^2} = r^2 e^{rx}$ and for the equation $\dfrac{d^2y}{dx^2} = -a^2y$: $r^2 e^{rx} = -a^2 e^{rx}$. This furnishes the auxiliary equation $r^2 = -a^2$ the roots of which are $r = \pm ai$. From the roots we get the general solution $y = C_1 e^{aix} + C_2 e^{-aix}$ to which the form can be given: $y = C_1 \cos ax + C_2 \sin ax$ by applying the relation $e^{ix} = \cos x + i \sin x$ which in itself represents an outstanding link between e and the trigonometric functions.

From $e^{ix} = \cos x + i \sin x$ and $e^{-ix} = \cos x - i \sin x$ we obtain, through addition

$$\cos x = \frac{e^{ix} + e^{-ix}}{2}$$ and through subtraction

$$\sin x = \frac{e^{ix} - e^{-ix}}{2i} .$$ In analogy to these functions of x the "hyperbolic" functions were defined as $\cosh x = \dfrac{e^x + e^{-x}}{2}$ and $\sinh x = \dfrac{e^x - e^{-x}}{2}$ which in turn led to the Gudermannian function $gdx \equiv \arctan$ (sinh x) with its application to the Mercator chart of the earth etc.; all are based upon the constant e.

Since our oldest records in the history of mathematics, special attention had been paid to the sequence: $1; 2; 4; 8; 16; 32; 64;$ $128; \cdots$ obtained through continued "duplication" which, up to the sixteenth century, had been considered as a special operation besides multiplication. The differences between the consecutive terms of the sequence $1; 2; 4; 8; \cdots$ and their differences reproduce the same sequence:

$$
\begin{array}{l}
1 \\
\quad > 1 \\
2 \\
\quad > 1 \\
\quad\quad > 2 \\
4 \quad\quad > 1 \\
\quad > 2 \\
\quad\quad > 2 \quad > 1 \\
\quad\quad\quad > 4 \\
8 \quad\quad > 2 \quad > 1 \\
\quad > 4 \quad\quad > 2 \\
\quad\quad > 4 \quad > 2 \quad > 1 \\
16 \quad > 8 \\
\quad\quad > 8 \quad > 4 \\
\quad > 16 \quad > 8 \\
32 \quad\quad > 16 \\
\quad > 32 \\
64
\end{array}
$$

$\cdot \; \cdot \; \cdot \; \cdot \; \cdot \; \cdot \; \cdot \; \cdot \; \cdot \; \cdot$

If x stands for any positive integer we have $\dfrac{2^{x+1}-2^x}{(x+1)-x}=2^x$. Using the notation $\Delta 2^x \equiv 2^{x+1}-2^x$ and $\Delta x \equiv (x+1)-x$ we get $\dfrac{\Delta 2^x}{\Delta x}=2^x$ in analogy to $\dfrac{de^x}{dx}=e^x$. The quality of reproducing itself which the exponential function $y=2^x$ has in regard to its differences when the exponent increases from 1 to 2 to 3 $\cdots$ reappears for the function $y=e^x$ in regard to the derivatives.

The relationship between the functions $y=2^x$ and $y=e^x$ can be illustrated through the following example in reference to a banking account. If a bank were to operate on a 100% compound interest rate, thus adding to the invested capital its own value as its interest after one year the investment of one dollar would be raised to two dollars at the end of the first year, to four dollars at the end of the second, to eight at the end of the third, etc. After n years we come to 2^n. If the bank were to use semi-annual compounding one-half of a year's interest would be added after the first half year. With the 100% interest rate, one-half dollar would be added after the first half year. During the second half of the year the capital would be $1+\frac{1}{2}$ and adding half of its value at the end of the second half year would bring it up to

$$(1+\tfrac{1}{2})+\tfrac{1}{2}(1+\tfrac{1}{2})=(1+\tfrac{1}{2})^2=1.5^2=2.25.$$

The original investment of one dollar will thus be raised to two dollars and twenty-five cents. Were the bank adding the interest to the capital after every one-third of a year, we would get at the end of the first year:

$$[(1+\tfrac{1}{3})+\tfrac{1}{3}(1+\tfrac{1}{3})]+\tfrac{1}{3}[(1+\tfrac{1}{3})+\tfrac{1}{3}(1+\tfrac{1}{3})]$$
$$=(1+\tfrac{1}{3})^3=2.37\cdots.$$

Similarly compounding after every quarter of the year would lead to

$$(1+\tfrac{1}{4})^4=2.44\cdots.$$

In general compounding n times per year, we would get from one dollar, at the end of the year $\left(1+\dfrac{1}{n}\right)^n$ and if n approaches infinity we come to $\lim\limits_{n\to\infty}\left(1+\dfrac{1}{n}\right)^n$ which is e. After x years with annual compounding we would obtain 2^x with continued compounding e^x. The process illustrated in reference to a bank account can be applied to any form of a growth and e can thus be considered as the value of 2 increased through a transformation of the repeated duplication into a continued process of growth.

If we then consider an entity a involved in a growth in which an increment from step to step is proportional to the value reached at the respective steps, k being a constant factor, we obtain:

beginning	a	
after the first step	$a+ka$	$=a(1+k)$
after the second step	$a(1+k)\quad+ka(1+k)$	$=a(1+k)^2$
after the third step	$a(1+k)^2\quad+ka(1+k)^2$	$=a(1+k)^3$
$\cdot\;\cdot\;\cdot\;\cdot$		
after the nth step	$a(1+k)^{n-1}+ka(1+k)^{n-1}=a(1+k)^n.$	

The result is a geometric progression with the ratio $1+k$. Almost any natural growth follows this same principle. The larger the population of a town has become, the larger will be the additional population per annum (under normal conditions), the larger an institution has become, the larger will be its further growth if the conditions continue which prevailed during its development. The closest possible relation between the increment and the growing entity prevails when $k=1$, that is when the increment equals the growing entity. In this case we get a geometric progression with the ratio 2, and if the process develops into a continued growth, 2 is replaced by e. The term "natural" in regard to e is justified.

The relationship between e and the sequence $1; 2; 4; 8; \cdots$ comes to the fore in the development of e as a continued fraction. Whereas e written as a decimal fraction shows no regularity in the sequence of its digits, it does present as a continued fraction the following regular form:

dicular has been added to the following radius etc. The consecutive footpoints lie on a logarithmic spiral. The ratios of the distances from the center of the circle to

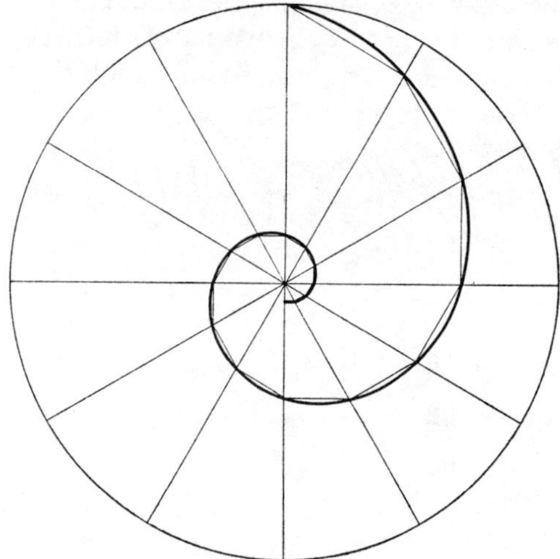

Fig. 1. Construction of a logarithmic spiral.

the points of the logarithmic spiral taken along two consecutive radii are

$$\cos \frac{360°}{12} = \cos 30°$$

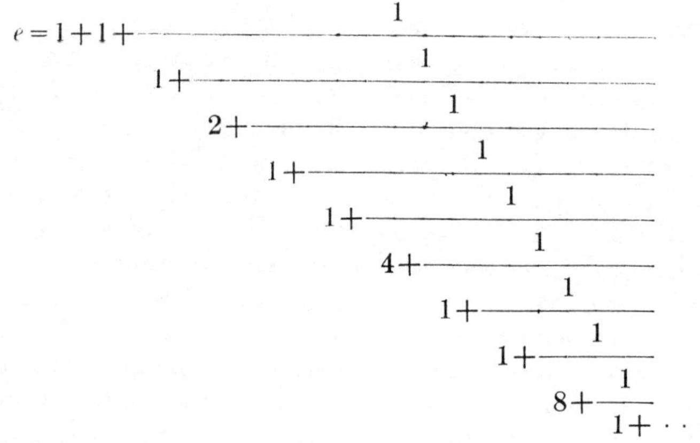

$$e = 1 + 1 + \cfrac{1}{1 + \cfrac{1}{2 + \cfrac{1}{1 + \cfrac{1}{1 + \cfrac{1}{4 + \cfrac{1}{1 + \cfrac{1}{1 + \cfrac{1}{8 + \cfrac{1}{1 + \cdots}}}}}}}}}$$

e can be derived also from a geometric viewpoint. In Figure 1 a circle has been divided into twelve equal parts. Each point of division has been joined with the center of the circle so that twelve radii were obtained. From the end of one of the radii (the one vertically upwards) a perpendicular has been drawn to the next radius. From its footpoint another perpen-

If r stands for the radius of a circle, the logarithmic spiral will arrive after its radius has turned 360° around the center, at a distance of $r \cdot (\cos 30°)^{12} = r \cdot 0.1778$ from the center. For every choice made in respect to the amount of points of division along the circle the construction will lead to a particular form of a logarithmic spiral. These spirals move faster to the center if there are fewer points of· division and slower if there are more.

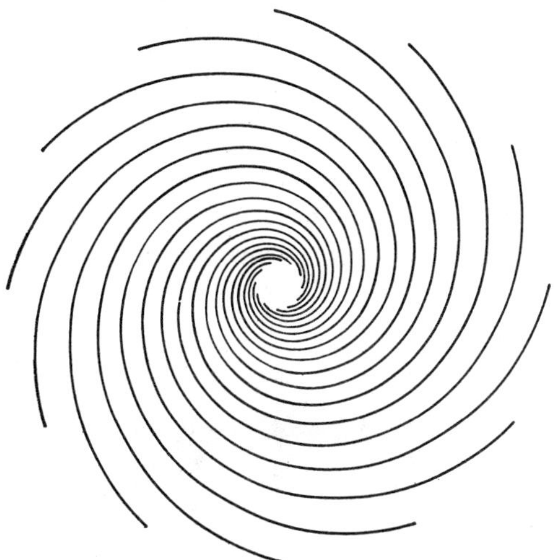

FIG. 2. Family of logarithmic spirals.

In Figure 2 logarithmic spirals have been drawn along the same lines as in Figure 1 but here at every one of the twelve points of division along the circle, a logarithmic spiral has been started; the construction lines are omitted.

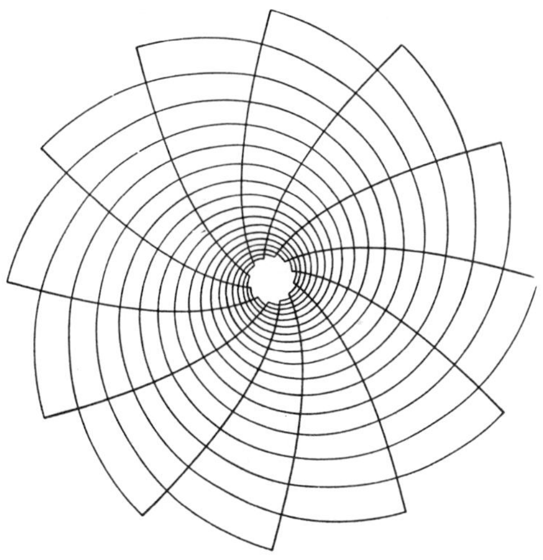

FIG. 3. Family of logarithmic spirals and orthogonal trajectories.

Figure 3 contains the same family of logarithmic spirals as Figure 2, but to them have been added the orthogonal trajectories (curves intersecting the family of curves at right angles). Those are again logarithmic spirals but of a different form.

They move very much faster towards the center than the spirals of the original family. Changing of the original family of logarithmic spirals, by changing in their construction the number of points of division along the circle would also change the orthogonal trajectories but in an opposite manner. If the original spirals were made to lead more straightly at the center the orthogonal trajectories would become rounder and vice versa. Therefore one particular family of logarithmic spirals will exist which will have the same form as their orthogonal trajectories. This particular family of logarithmic spirals is shown in Figure 4.

FIG. 4. Logarithmic spirals which are of the same form as their orthogonal trajectories.

The equation of this particular family of logarithmic spirals has to satisfy the condition that its basic ratio (between radii forming a given angle) is the same as the one of the orthogonal trajectories, both families considered rotating in opposite directions. In polar coordinates the equation of a logarithmic spiral with any ratio a between two radii forming an angle of one radian (57.29°, the angle for which the arc equals the radius) is $\rho = a^{\theta+c}$, ρ being the radius vector, θ the vectorial angle measured in radians and c the arbitrary parameter of the family of curves determining the position of a single spiral in regard to the polar axis. Differentiating $\rho = a^{\theta+c}$ to come to the differential equation of the family of spirals gives $\dfrac{d\rho}{d\theta} = a^{\theta+c} \cdot \ln a$. Eliminating c by substituting $a^{\theta+c} = \rho$ gives $\dfrac{d\rho}{d\theta} = \rho \cdot \ln a$. In order to come to the differential equation of the family of orthogonal trajectories $\dfrac{d\rho}{d\theta}$ is replaced by $-\rho^2 \cdot \dfrac{d\theta}{d\rho}$ and we obtain $-\rho^2 \cdot \dfrac{d\theta}{d\rho} = \rho \cdot \ln a$ or $-\rho \dfrac{d\theta}{d\rho} = \ln a$. Solving the differential equation by separating the variables we obtain

$$-d\theta = \ln a \cdot \frac{d\rho}{\rho}$$

and through integration:

$$-\theta = \ln a \cdot \ln \rho + c_1;$$

c_1 being the constant of integration. Solving for ρ we get

$$\ln a \cdot \ln \rho = -\theta - c_1$$

$$\ln \rho = \frac{1}{\ln a}(-\theta - c_1)$$

$$\rho = e^{1/\ln a \cdot (-\theta - c_1)} = e^{-1/(\ln a)\theta - c_1/(\ln a)}$$

and for $e^{-c_1/\ln a} \equiv c_2$, we obtain

$$\rho = e^{-1/(\ln a)\theta + c_2}.$$

The result is the equation of another family of logarithmic spirals. Writing for the equation of the original spirals instead of $\rho = a^{\theta+c}$ the expression

$$\rho = (e^{\ln a})^{\theta+c} = e^{\ln a \cdot \theta + \ln a \cdot c}$$

and for $e^{\ln a \cdot c} \equiv c_3$ we get $\rho = e^{\ln a \theta + c_3}$. The condition that both families have the same ratio but turn in the opposite sense (negative sign for θ) is $\ln a = \dfrac{1}{\ln a}$ or $(\ln a)^2 = 1$; $\ln a = \pm 1$. Therefore $e^{\ln a} = e^{\pm 1}$ and $a = e^{\pm 1}$. The first solution is $a = e$ and the second $a = \dfrac{1}{e}$. Therefore the particular family of logarithmic spirals has the equation $\rho = e^{\theta+c_3}$ or $\rho = e^{-\theta+c_3}$. For the corresponding

families of orthogonal trajectories we obtain by substituting the values $\ln a = \pm 1$:

$$\rho = e^{-\theta+c_2} \quad \text{and} \quad \rho = e^{\theta+c_2}.$$

The two solutions express a reciprocal relationship between the two families of logarithmic spirals; if one represents the original family the other represents the orthogonal trajectories and vice versa. We therefore have only one pair of families of logarithmic spirals which are orthogonal trajectories.

Fig. 5

Fig. 6

In the equation of the logarithmic spirals of Figure 4: $\rho = e^{\theta+c_3}$ and $\rho = e^{-\theta+c_3}$ e takes the place of a. If we take two radii separated by an angle of 1 radian their lengths are $\rho_1 = e^{\theta + c_3}$ and $\rho_2 = e^{\theta + 1 + c_3}$ and their ratio is

$$\frac{\rho_2}{\rho_1} = \frac{e^{\theta + 1 + c_3}}{e^{\theta + c_3}} = e.$$

The particular system of logarithmic spirals therefore contains the ratio *e* between any two radii which form an angle of one radian.

Logarithmic spirals being an adequate expression of an organic growth, of an evolutionary principle, have inspired the great mathematician Jacques Bernoulli to order a logarithmic spiral to be engraved on his tombstone with the inscription: "Eadem mutato resurgo (In different form, yet still the same, shall I arise)." Among all possible logarithmic spirals only one form has the quality of balance to be congruent to its orthogonal trajectories and this one is based on the ratio *e*. Taken as a form this logarithmic spiral expresses a specific harmony.

The three figures 5–7 (pp. 354–355) contain the same families of curves as Figure 4, the logarithmic spirals based upon *e*, but they emphasize different elements in order to bring out various characteristics. In Figure 5 the areas between the logarithmic spirals are shown alternately in black and

white in a checkerboard manner. In Figure 6 the areas between one system of logarithmic spirals are alternately black and white which underlines the spiral form. In Figure 7 the distribution of the

Fig. 7

black and white areas forms an axis or symmetry and brings out the mutual relation of the orthogonal trajectories. The diagrams presenting the outstanding systems of logarithmic spirals based upon *e* may also be considered among the many contributions of mathematics to the domain of the arts.

The evolution of extended decimal approximations to π

by J. W. Wrench, Jr., Applied Mathematics Laboratory,
David Taylor Model Basin, Washington, D.C.

In his historical survey of the classic problem of "squaring the circle," Professor E. W. Hobson [1]* distinguished three distinct periods, characterized by fundamental differences in method, immediate aims, and available mathematical tools.

The first period—the so-called geometrical period—extended from the earliest empirical determinations of the ratio of the circumference of a circle to its diameter to the invention of the calculus about the middle of the seventeenth century. The main effort was directed toward the approximation of this ratio by the calculation of perimeters or areas of regular inscribed and circumscribed polygons.

The second period began in the middle of the seventeenth century and lasted for more than a hundred years. During this period the methods of the calculus were employed in the development of analytical expressions for π in the form of infinite series, products, and continued fractions.

The third period, which extended from the middle of the eighteenth century to nearly the end of the nineteenth century, was devoted to studies of the nature of the number π. J. H. Lambert [2] proved the irrationality of π in 1761, and F. Lindemann [3] first established its transcendence in 1882.

This article is concerned with the second period and its sequel, which extends to the present day.

According to Hobson [1], the first analytical expression discovered in this period is the infinite product

$$\frac{\pi}{2} = \frac{2}{1} \cdot \frac{2}{3} \cdot \frac{4}{3} \cdot \frac{4}{5} \cdot \frac{6}{5} \cdot \frac{6}{7} \cdot \frac{8}{7} \cdot \frac{8}{9} \cdots,$$

which was published by John Wallis [4] in 1655.

Lord Brouncker, the first president of the Royal Society, about 1658 found the infinite continued fraction

$$\frac{\pi}{4} = \frac{1}{1+} \frac{1^2}{2+} \frac{3^2}{2+} \frac{5^2}{2+} \cdots,$$

which was shown subsequently by Euler to be equivalent to the alternating series

$$\frac{\pi}{4} = 1 - \frac{1}{3} + \frac{1}{5} - \frac{1}{7} + \frac{1}{9} - \cdots,$$

known to G. W. Leibniz in 1674.

The great majority of calculations of π to many decimal places have been based upon the power series

$$\arctan x = x - \frac{x^3}{3} + \frac{x^5}{5} - \cdots, \quad -1 \leq x \leq 1,$$

which was discovered in 1671 by James Gregory [5]. He failed, however, to note explicitly the special case corresponding to $x = 1$, which is ascribed to Leibniz.

* Numbers in brackets refer to the references listed at the end of the article.

Sir Isaac Newton [6] in 1676 discovered the power series

$$\arcsin x = x + \frac{1}{2}\,\frac{x^3}{3} + \frac{1\cdot 3}{2\cdot 4}\,\frac{x^5}{5} + \cdots ,$$

$$-1 \leqq x \leqq 1,$$

which has been used by a few computers of π.

In 1755 Leonhard Euler [7] obtained the following useful series:

$$\arctan x = \frac{x}{1+x^2}\left\{ 1 + \frac{2}{3}\left(\frac{x^2}{1+x^2}\right) \right.$$
$$\left. + \frac{2\cdot 4}{3\cdot 5}\left(\frac{x^2}{1+x^2}\right)^2 + \cdots \right\}.$$

It was by means of Gregory's series, taking $x = 1/\sqrt{3}$, that Abraham Sharp [8], at the suggestion of the English astronomer Edmund Halley, computed π to 72 decimal places in 1699, thereby nearly doubling the greatest accuracy (39 decimal places) attained by earlier computers, who had used geometrical methods. Sharp's calculation was extended by Fautet de Lagny [9] in 1719 to 127 decimals (the 113th place has a unit error).

Newton set $x = \frac{1}{2}$ in his series, and thereby computed π to 14 places. A Japanese computer, Matsunaga Ryohitsu [10], used the same procedure to evaluate π correct to 49 decimal places in 1739. About 1800 a Chinese, Chu Hung, calculated π to 40 places (25 correct) by this series [10].

Most computers of π in modern times have used Gregory's series in conjunction with certain arctangent relations. Only nine of these relations have been employed to any extent in such computations. We shall now consider these formulas, arranged according to the increasing precision of the approximations computed by their use.

$$\text{I.} \quad \frac{\pi}{4} = 5 \arctan \frac{1}{7} + 2 \arctan \frac{3}{79}$$

Euler [7] in 1755 used this relation in conjunction with his series for arctan x to

compute π correct to 20 decimal places in one hour. Baron Georg von Vega [11] in 1794 employed Gregory's series and the preceding relation to evaluate π to 140 decimal places, of which the first 136 were correct. This precision was exceeded by that attained by an unknown calculator whose manuscript, containing an approximation correct to 152 places, was seen in the Radcliffe Library at Oxford toward the close of the eighteenth century.

$$\text{II.} \quad \frac{\pi}{4} = 4 \arctan \frac{1}{5} - \arctan \frac{1}{70} + \arctan \frac{1}{99}$$

Euler published this relation in 1764. It was used by William Rutherford [12] in 1841 to compute π to 208 places (152 correct).

$$\text{III.} \quad \frac{\pi}{4} = \arctan \frac{1}{2} + \arctan \frac{1}{5} + \arctan \frac{1}{8}$$

This formula was supplied the calculating prodigy Zacharias Dahse [13] by L. K. Schulz von Strassnitzky of Vienna. Within a period of two months in 1844, Dahse thereby evaluated π correct to 200 places.

$$\text{IV.} \quad \frac{\pi}{4} = \arctan \frac{1}{2} + \arctan \frac{1}{3}$$

First published by Charles Hutton [14] in 1776, this relation was used by W. Lehmann [15] of Potsdam to compute π to 261 decimals in 1853. Tseng Chi-hung [16] in 1877 used the same formula to evaluate π to 100 decimals in a little more than a month.

$$\text{V.} \quad \frac{\pi}{4} = 2 \arctan \frac{1}{3} + \arctan \frac{1}{7}$$

The relation was also published by Hutton [14] in 1776, and independently by Euler in 1779. Vega [17] used it in 1789 to compute 143 decimals (126 correct). In order to remove the uncertainty caused by the discrepant approximations of Rutherford and Dahse, Thomas Clausen [18] extended the calculation to 248 correct decimals in 1847, and Lehman [15] reached 261 decimals in 1853 by this formula, confirming his independent calculation of π to the same extent by relation IV. Edgar Frisby [19] in Washington, D. C. used relation V in conjunction with Euler's series to compute π to 30 places in 1872.

$$\text{VI.}\quad \frac{\pi}{4}=3 \arctan \frac{1}{4}+\arctan \frac{1}{20}+\arctan \frac{1}{1985}$$

This formula was published by S. L. Loney [20] in 1893, by Carl Störmer [21] in 1896, and was rediscovered by R. W. Morris [22] in 1944. By means of this formula D. F. Ferguson, then of the Royal Naval College, Eaton, Chester, England, performed a longhand calculation of π to 530 decimal places between May 1944 and May 1945. At that time he discovered a discrepancy between his approximation and the final result of William Shanks—discussed under formula IX—beginning with the 528th place. The first notice of an error in Shanks's well-known approximation appeared in a note [22] published by Ferguson in March 1946. He continued his calculation of π and in July 1946 published [23] a correction to Shanks's value through the 620th decimal place. Subsequently, Ferguson used a desk calculator to reach 710 decimals [24] by January 1947, and finally 808 decimals [25] by September 1947.

$$\text{VII.}\quad \frac{\pi}{4}=8 \arctan \frac{1}{10}-\arctan \frac{1}{239}$$
$$-4 \arctan \frac{1}{515}$$

S. Klingenstierna discovered this relation in 1730; it was rediscovered more than a century later by Schellbach [26]. It was used by C. C. Camp [27] in 1926 to evaluate $\pi/4$ to 56 places. D. H. Lehmer [28] recommended it in conjunction with the next formula for the calculation of π to many figures. G. E. Felton on March 31, 1957 completed a calculation of π to 10021 places on a Pegasus computer at the Ferranti Computer Centre in London.

This required 33 hours of computer time. The result was published to 10000 places [29]. A check calculation using formula VIII revealed that, because of a machine error, this result was incorrect after 7480 decimal places.

Gauss [30] investigated the derivation of arctangent relations and reduced it to a problem in Diophantine analysis. Relation VIII is one of several formulas he developed. J. P. Ballantine [31] substantiated Lehmer's claim that this formula is especially effective for extensive calculation, by discussing its use in conjunction with Euler's series for the arctangent.

$$\text{VIII.}\quad \frac{\pi}{4}=12 \arctan \frac{1}{18}+8 \arctan \frac{1}{57}$$
$$-5 \arctan \frac{1}{239}$$

Felton carried out a second calculation to 10021 places, and by March 1, 1958 had removed all discrepancies from his results, so that the approximations computed from formulas VII and VIII agreed to within 3 units in the 10021st decimal place. The corrected result remains unpublished.

$$\text{IX.}\quad \frac{\pi}{4}=4 \arctan \frac{1}{5}-\arctan \frac{1}{239}$$

This is the most celebrated of all the relations of this kind. John Machin, its discoverer, computed π correct to 100 decimals by means of it in conjunction with Gregory's series, and the result [32] appeared in 1706. Clausen [18] in 1847 used this relation in addition to Hutton's formula V to compute π correct to 248 decimal places, as has already been noted.

Rutherford resumed his calculation of π in 1852, using Machin's formula this time, as did his former pupil William Shanks. Shanks's first published approximation to π contained 530 decimal places, and was incorporated in Rutherford's note [33], published in 1853, which set forth his approximation to 441 decimals.

Later that year Shanks published his book [34] containing an approximation to 607 places and giving all details of the calculation to 530 places. It is now known that Shanks's value was incorrectly calculated beyond 527 decimal places. The accuracy of that value was further vitiated by a blunder committed by Shanks in correcting his copy prior to publication, with the result that similar errors appear in decimal places 460–462 and 513–515. These errors persist in Shanks's first paper of 1873 [35] containing the extension to 707 decimals of his earlier approximation. His second paper of that year [36], which contained his final approximation to π, gives corrections of these errors; however, there appears an inadvertent typographical error in the 326th decimal place of his final value. In retrospect, we now realize that Shanks's first value published in 1853 was the most accurate he ever published.

The accuracy of Shanks's approximation to at least 500 decimals was confirmed by the independent calculations of Professor Richter [37] of Elbing, Germany, who in 1853–1854 computed successive approximations to 330, 400, and 500 places. Richter's communications do not reveal the formula that he used.

Machin's formula was used by H. S. Uhler in an unpublished computation correct to 282 places, which was completed in August 1900.

F. J. Duarte computed π correct to 200 places by this method in 1902. The result was published [38] six years later.

As a by-product of his calculation of the natural logarithms of small primes, Uhler in 1940 noted [39] confirmation to 333 decimal places of Shanks's approximation.

In December 1945, Professor R. C. Archibald suggested that the writer undertake the computation of π by Machin's formula in order to provide an independent check of the accuracy of Ferguson's calculations. With the collaboration of Levi B. Smith, who evaluated arctan 1/239 to 820 decimal places, the writer computed π to 818 places by February 1947, using a desk calculator. The result was published [24] to 808 places in April 1947, and was verified to 710 places by Ferguson in a note published concurrently [24]. The limit of 808 decimals in the published value was chosen to provide precision comparable to that obtained by P. Pedersen [40] in his approximation to e.

Collation of this 808-place approximation with results obtained by Ferguson later that year revealed several erroneous figures beyond the 723rd place in the writer's approximation to arctan $\frac{1}{5}$. These errors vitiated the corresponding figures in the approximation to π. Corrections of these errors and extensions of Ferguson's results appeared in a joint paper [25] by Ferguson and the writer in January 1948, which concluded with an 808-place approximation to π of guaranteed accuracy.

Subsequently, Smith and the writer resumed their calculations and by June 1949 had obtained an approximation to about 1120 decimal places [41]. Before final checking of this extension could be completed, the ENIAC (Electronic Numerical Integrator and Computer) at the Ballistic Research Laboratories, Aberdeen Proving Ground, was employed by George W. Reitwiesner and his associates in September 1949 to evaluate π to about 2037 places (2040 working decimals) in a total time (including card handling) of 70 hours [42]. Machin's formula was also used in this computation.

In November 1954, Smith and the writer extended their calculation to 1150 places, and in January 1956 reverted to this work once more to attain their final result, which was terminated at 1160 places, of which the first 1157 agree with those obtained on the ENIAC.

A calculation of π was performed in duplicate on the NORC (Naval Ordnance Research Calculator) in November 1954 and in January 1955 as a demonstration problem, prior to the delivery of that computer to the U. S. Naval Proving

Grounds at Dahlgren, Virginia. Again, Machin's formula was selected, and the calculation was completed to 3093 decimal places in 13 minutes running time. A report of this work, in which the value of π was presented unrounded to 3089 decimal places, was published by S. C. Nicholson and J. Jeenel [43] of the Watson Scientific Computing Laboratory, in New York.

In January 1958, François Genuys [44] programmed and carried out the evaluation of π correct to 10000 decimal places on an IBM 704 Electronic Data Processing System at the Paris Data Processing Center. Machin's formula in conjunction with Gregory's series was used. Only 40 seconds were required to attain the 707 decimal-place precision reached by Shanks, and one hour and forty minutes was required to reach the 10000 places of the final result.

On July 20, 1959, the program of Genuys was used on an IBM 704 system at the Commissariat à l'Energie Atomique in Paris to compute π to 16167 decimal places. This latest approximation is unpublished at present.

The motivation of modern calculations of π to many decimal places was conjectured by Professor P. S. Jones [45] in 1950 as being attributable to "intellectual curiosity and the challenge of an unchecked and long untouched computation." This reason for undertaking such work should be supplemented by reference to the recurrent interest in determining a statistical measure of the randomness of distribution of the digits in the decimal representation of π.

Augustus De Morgan [46] drew attention to the deficiency in the number of appearances of the digit 7 in Shanks's 607-place approximation to π. In 1897 E. B. Escott [47] raised the question whether the deficiency of 7's noted in Shanks's final approximation could be explained.

In June 1949, the late Professor John von Neumann expressed an interest in utilizing the ENIAC to determine the value of π and e to many places as the basis for a statistical study of the distribution of their decimal digits. A statistical treatment of the first 2000 decimal digits of both π and e was published by N. C. Metropolis, G. Reitwiesner, and J. von Neumann [48]. Further analysis of these data was performed by R. E. Greenwood [49], using the coupon collector's test. A

TABLE 1

CUMULATIVE DISTRIBUTION OF THE FIRST 16000 DECIMAL DIGITS OF π

THOUSAND	DIGIT									
	0	1	2	3	4	5	6	7	8	9
1	93	116	103	102	93	97	94	95	101	106
2	182	212	207	188	195	205	200	197	202	212
3	259	309	303	265	318	315	302	287	310	332
4	362	429	408	368	405	417	398	377	405	431
5	466	532	496	459	508	525	513	488	492	521
6	557	626	594	572	613	622	619	606	582	609
7	657	733	692	686	702	730	708	694	680	718
8	754	833	811	781	809	834	816	786	764	812
9	855	936	911	884	910	933	914	883	854	920
10	968	1026	1021	974	1012	1046	1021	970	948	1014
11	1070	1099	1111	1080	1133	1150	1129	1070	1031	1127
12	1162	1193	1214	1176	1233	1262	1227	1166	1144	1223
13	1266	1314	1316	1272	1343	1358	1324	1260	1246	1301
14	1365	1416	1419	1383	1440	1455	1426	1344	1339	1413
15	1456	1513	1511	1491	1553	1549	1520	1441	1458	1508
16	1556	1601	1593	1602	1670	1659	1615	1548	1546	1610

count of each of the decimal digits appearing in the NORC approximation appears in the paper of Nicholson and Jeenel [43]. A number of recent investigators have discussed the distribution of digits in Shanks's approximation and in the corrected value of π. These investigators include F. Bukovszky [50], W. Hope-Jones [51], E. H. Neville [52], and B. C. Brookes [53].

The writer has recently completed a count by centuries of the 16167 decimal digits constituting the fractional part of the latest approximation to π. An abridgment of this information is presented in the accompanying table.

The standard χ^2 test for goodness of fit reveals no abnormal behavior in the distribution of digits in this sample; in particular, there appears to be no basis for supposing that π is not simply normal [54] in the decimal scale of notation. It has been pointed out recently by Ivan Niven [55] that the normality of such numbers as π, e, and $\sqrt{2}$ has yet to be proved.

Numerical studies directed toward the empirical investigation of the normality of π clearly require increasingly higher decimal approximations, which can best be obtained by use of ultra-high-speed electronic computers now under design and development.

References

1. E. W. Hobson, "*Squaring the Circle,*" a History of the Problem (Cambridge, 1913; reprinted by Chelsea Publishing Company, New York, 1953).
2. J. H. Lambert, "Mémoire sur quelques propriétés rémarquables des quantités transcendentes circulaires et logarithmiques," *Histoire de l'Académie de Berlin, 1761* (1768).
3. F. Lindemann, "Ueber die Zahl π," *Mathematische Annalen*, 20 (1882), 213–225.
4. J. Wallis, *Arithmetica Infinitorum* (1655).
5. R. C. Archibald, *Outline of the History of Mathematics* (6th ed.; Buffalo, N. Y.: The Mathematical Association of America, 1949), p. 40.
6. Letter from Newton to Oldenburg dated October 24, 1676.
7. E. Beutel, *Die Quadratur des Kreises* (Leipzig, 1920), p. 40. See also E. W. Hobson,[1] p. 39.
8. H. Sherwin, *Mathematical Tables* (London, 1705), p. 59.
9. F. de Lagny, "Mémoire Sur la Quadrature du Circle, & sur la mesure de tout Arc, tout Secteur, & tout Segment donné," *Histoire de l'Académie Royale des Sciences, 1719* (Paris, 1721), pp. 135–145.
10. Y. Mikami, *The Development of Mathematics in China and Japan* (Leipzig, 1913), p. 202 and p. 141.
11. G. Vega, *Thesaurus Logarithmorum Completus* (Leipzig, 1794; reprinted by G. E. Stechert & Co., New York, 1946), p. 633.
12. W. Rutherford, "Computation of the Ratio of the Diameter of a Circle to its Circumference to 208 places of figures," *Philosophical Transactions of the Royal Society of London*, 131 (1841), 281–283.
13. Z. Dahse, "Der Kreis-Umfang für den Durchmesser 1 auf 200 Decimalstellen berechnet," *Journal für die reine und angewandte Mathematik*, 27 (1844), 198.
14. *Philosophical Transactions of the Royal Society of London*, 46 (1776), 476–492.
15. W. Lehmann, "Beitrag zur Berechnung der Zahl π, welche das Verhältniss des Kreis-Durchmessers zum Umfang ausdrückt," *Archiv der Mathematik und Physik*, 21 (1853), 121–174.
16. Y. Mikami,[10] pp. 141–142.
17. *Nova Acta Academiae Scientiarum Imperialis Petropolitanae*, 9 (1790), 41.
18. *Astronomische Nachrichten*, 25 (1847), col. 207–210.
19. E. Frisby, "On the calculation of π," *Messenger of Mathematics*, 2 (1873), 114–118.
20. S. L. Loney, *Plane Trigonometry* (Cambridge, 1893), p. 277.
21. C. Störmer, "Sur l'application de la théorie des nombres entiers complexes à la solution en nombres rationnels x_1, x_2, $\cdots$, x_n, c_1, c_2, $\cdots$, c_n, k de l'équation $c_1 \operatorname{arctg} x_1 + c_2 \operatorname{arctg} x_2 + \cdots + c_n \operatorname{arctg} x_n = k\pi/4$," *Archiv for Mathematik og Naturvidenskab*, 19 (1896), 70.
22. D. F. Ferguson, "Value of π," *Nature*, 157 (1946), 342. See also D. F. Ferguson, "Evaluation of π. Are Shanks' figures correct?" *Mathematical Gazette*, 30 (1946), 89–90.
23. R. C. Archibald, "Approximations to π," *Mathematical Tables and other Aids to Computation*, 2 (1946-1947), 143–145.
24. L. B. Smith, J. W. Wrench, Jr., and D. F. Ferguson, "A New Approximation to π," *ibid.*, 2 (1946–1947), 245–248.
25. D. F. Ferguson and J. W. Wrench, Jr., "A New Approximation to π (conclusion)," *ibid.*, 3 (1948-1949), 18–19. See also R. Liénard, "Constantes mathématiques et système binaire," *Intermédiaire des Recherches Mathématiques*, 5 (1948), 75.

26. K. H. Schellbach, "Über den Ausdruck $\pi = (2/i)\log i$," *Journal für die reine und angewandte Mathematik*, 9 (1832), 404–406.

27. C. C. Camp, "A New Calculation of π," *American Mathematical Monthly*, 33 (1926), 474.

28. D. H. Lehmer, "On Arccotangent Relations for π," *ibid.*, 45 (1938), 657–664.

29. G. E. Felton, "Electronic Computers and Mathematicians," *Abbreviated Proceedings of the Oxford Mathematical Conference for Schoolteachers and Industrialists at Trinity College, Oxford, April 8–18, 1957*, p. 12–17; footnote, p. 12–53.

30. C. F. Gauss, *Werke* (Göttingen, 1863; 2nd ed., 1876), Vol. 2, p. 499–502.

31. J. P. Ballantine, "The Best (?) Formula for Computing π to a Thousand Places," *American Mathematical Monthly*, 46 (1939), 499–501.

32. W. Jones, *Synopsis palmiorum matheseos* (London, 1706), p. 263.

33. W. Rutherford, "On the Extension of the value of the ratio of the Circumference of a circle to its Diameter," *Proceedings of the Royal Society of London*, 6 (1850–1854), 273–275. See also *Nouvelles Annales des Mathématiques*, 14 (1855), 209–210.

34. W. Shanks, *Contributions to Mathematics, comprising chiefly the Rectification of the Circle to 607 Places of Decimals* (London, 1853).

35. W. Shanks, "On the Extension of the Numerical Value of π," *Proceedings of the Royal Society of London*, 21 (1873), 318.

36. W. Shanks, "On certain Discrepancies in the published numerical value of π," *ibid.*, 22 (1873), 45–46.

37. *Archiv der Mathematik und Physik*, 21 (1853), 119; 22 (1854), 473; 23 (1854), 475–476; 25 (1855), 471–472 (posthumous). See also *Nouvelles Annales des Mathématiques*, 13 (1854), 418–423.

38. *Comptes Rendus de l'Académie des Sciences de Paris*, Vol. 146, 1908. See also *L'Intermédiaire des Mathématiciens*, 27 (1920), 108–109, and F. J. Duarte, *Monografía sobre los Números π y e* (Caracas, 1949).

39. H. S. Uhler, "Recalculation and Extension of the Modulus and of the Logarithms of 2, 3, 5, 7, and 17," *Proceedings of the National Academy of Sciences*, 26 (1940), 205–212.

40. D. H. Lehmer, Review 275, *Mathematical Tables and other Aids to Computation*, 2 (1946–1947), 68–69.

41. J. W. Wrench, Jr., and L. B. Smith, "Values of the terms of the Gregory series for arccot 5 and arccot 239 to 1150 and 1120 decimal places, respectively," *ibid.*, 4 (1950), 160–161.

42. G. Reitwiesner, "An ENIAC Determination of π and e to more than 2000 Decimal Places," *ibid.*, 4 (1950), 11–15.

43. S. C. Nicholson and J. Jeenel, "Some Comments on a NORC Computation of π," *ibid.*, 9 (1955), 162–164.

44. F. Genuys, "Dix milles décimales de π," *Chiffres*, 1 (1958), 17–22.

45. P. S. Jones, "What's New About π?," THE MATHEMATICS TEACHER, 43 (1950), 120–122.

46. A. De Morgan, *A Budget of Paradoxes* (1st ed., 1872; 2nd ed., Chicago: The Open Court Publishing Company, 1915), Vol. 2, p. 65. See also James R. Newman, *The World of Mathematics* (New York: Simon and Schuster, 1956), Vol. 4, pp. 2379–2380.

47. E. B. Escott, Question 1154, *L'Intermédiaire des Mathématiciens*, 4 (1897), 221.

48. N. C. Metropolis, G. Reitwiesner, and J. von Neumann, "Statistical Treatment of the Values of First 2000 Decimal Digits of e and π Calculated on the ENIAC," *Mathematical Tables and other Aids to Computation*, 4 (1950), 109–111.

49. R. E. Greenwood, "Coupon Collector's Test for Random Digits," *ibid.*, 9 (1955), 1–5.

50. F. Bukovsky, "The Digits in the Decimal Form of π," *The Mathematical Gazette*, 33 (1949), 291.

51. W. Hope-Jones, "Surprising," *ibid.*, Vol. 35, 1951.

52. E. H. Neville, "The Digits in the Decimal Form of π," *ibid.*, 35 (1951), 44–45.

53. B. C. Brookes, "On the Decimal for π," *ibid.*, 36 (1952), 47–48.

54. G. H. Hardy and E. M. Wright, *An Introduction to the Theory of Numbers* (Oxford, 1938), pp. 123–127.

55. I. Niven, *Irrational Numbers*, Carus Monograph No. 11 (Buffalo, N. Y.: The Mathematical Association of America, 1956), p. 112.

THE NUMBER π *

By Herman Von Baravalle†

EDITOR'S NOTE.—The numbers π, e, and G (the ratio of the Golden Section) have held special fascination for many students of mathematics. The reader will find equally comprehensive and interesting treatments of the numbers e and G in Professor von Baravalle's earlier articles cited in the footnotes to the classic reproduced here.—F. JOE CROSSWHITE.

TWO outstanding constants of mathematics have been dealt with in previous articles in THE MATHEMATICS TEACHER, the number e, the base of the natural logarithms[1] and the number G, the ratio of the Golden Section.[2] To complete this series, the present article takes up the third and best known constant, the number π.

As its symbol indicates (π stands for periphery), it represents the ratio of the two outstanding dimensions of the circle, the way around it and the distance across it.

$$\pi = \frac{\text{circumference of a circle}}{\text{diameter of a circle}}.$$

Expressing the diameter in terms of the radius r, we obtain the formula for the circumference of the circle c:

$$\frac{c}{2r} = \pi; \qquad c = 2\pi r.$$

This is by far not the only ratio in which this constant appears. For instance, π is also the ratio of the area of a circle A to the area of the square erected on its radius r:

$$\pi = \frac{A}{r^2}; \qquad A = r^2\pi.$$

It further appears in many other formulae. The volume (V) of a circular cylinder with a base-radius r and altitude h is

$$V = r^2\pi h$$

and of a circular cone—

$$V = \frac{r^2\pi h}{3}.$$

The surface of a sphere is

$$A = 4r^2\pi$$

and its volume—

$$V = \frac{4}{3}r^3\pi.$$

The domain of π also extends beyond circular structures. The area of an ellipse with the semi-axes a and b is

$$A = ab\pi$$

and the volume of an ellipsoid with the three semi-axes a, b, and c is

$$V = \frac{4}{3}abc\pi.$$

The area enclosed in a cardioid drawn in

---

* Reprinted from THE MATHEMATICS TEACHER, XLV (May 1952), 340–48.

† At the time this article was published, the author was associated with Adelphi College, Garden City, New York.

[1] December 1945 issue (Volume XXXVIII, No. 8).

[2] January 1948 issue (Volume XLI, No. 1).

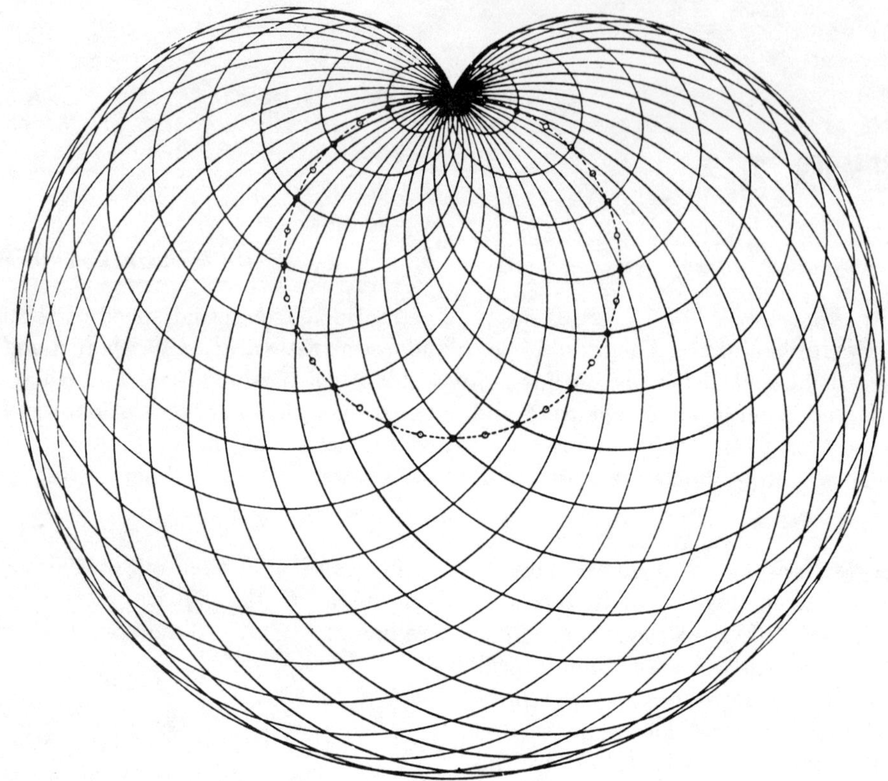

Fig. 1.—The Cardioid

Figure 1 as an envelope of circles is

$$A = \frac{3}{2} a^2 \pi,$$

in which a stands for the diameter of the circle, whose circumference is indicated by the dotted line.[3]

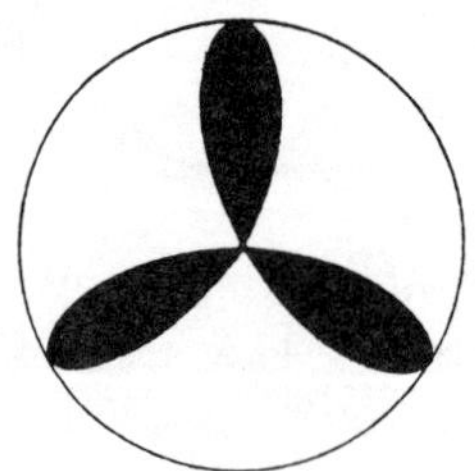

Fig. 2.—The Three-Leaved Rose

Further examples of curves whose formulae contain π are the roses. The area

enclosed by a three-leaved rose (black portions of Fig. 2) is

$$A = \frac{1}{4} a^2 \pi,$$

in which a stands for the radius of the circle circumscribed around it. The area

Fig. 3.—The Four-Leaved Rose

enclosed by a four-leaved rose (black area in Fig. 3) is

$$A = \frac{1}{2} a^2 \pi,$$

in which a again denotes the radius of the circumscribed circle. The volume of a ring

---

[3] To construct Figure 1, the dotted circle is divided into thirty-two equal parts. Each of the thirty-two points of division becomes the center of a circle whose radius is its distance from the highest point on the circle (upper end of vertical diameter).

(torus), obtained by rotating a circle with radius a about an axis in the same plane at a distance of b units from the center of the circle is expressed in the following formula:

$$V = 2a^2b\pi^2.$$

The volume of the solid of rotation produced by rotating an astroid about one of its axes is

$$V = \frac{32}{105}\,a^3\pi.$$

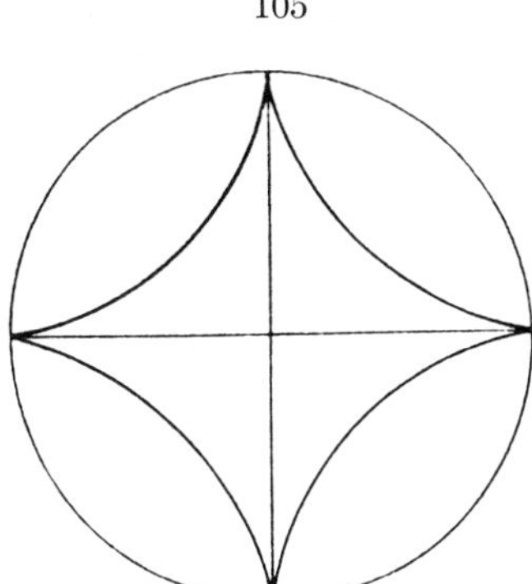

FIG. 4.—THE ASTROID

Here a represents the distance of any of the star points from the center, the radius of the circumscribed circle. The surface area of the same solid of rotation is

$$A = \frac{12}{5}\,a^2\pi.$$

The formula for the volume of the solid $x^{\frac{2}{3}}+y^{\frac{2}{3}}+z^{\frac{2}{3}}=a^{\frac{2}{3}}$ whose traces on the coordinate plane are astroids, also contains π:

$$V = \frac{4}{35}\,a^3\pi.$$

All these formulae are obtained by integral calculus.

We can also go beyond areas, surfaces, and volumes to find π again in a variety of relationships. A semicircle (radius r), cut out of sheet metal and balanced on a point, will be in equilibrium only when the point of support lies on its axis of symmetry at a distance d from the center, which is

$$d = \frac{4r}{3\pi}.$$

π even appears in formulae of probability, statistics, and in the field of an actuary.

Any vibration, mechanical, acoustical, or electrical, proceeds with varying speed. By determining the distance covered by a point on a vibrating musical chord between its extreme positions during a certain time unit, we obtain its average speed of motion. The actual speed of the point is greater every time the chord is near to passing its middle position. It is less than the average speed every time the point finds itself near to one of its extreme elongations. The maximum speed occurs when the point passes in either direction through its position of rest. This maximum speed is in any vibration exactly $\frac{\pi}{2}$ times the average speed.[4] As this holds good for vibrations accompanying every sound of our own vocal chords and in the air around us, π is contained in every word and sentence we say.

The value of π up to 22 decimal places is

$$3.1415926535897932384626\ldots.$$

These successive numerals are the same as

---

[4] The differential equation of a vibration is

$$\frac{d^2x}{dt^2} = -\,a^2x$$

and its complete solution is $x = c \sin(at - \alpha)$. In this equation, c and α represent arbitrary constants. For $x = 0$ at $t = 0$, the solution is

$$x = c \sin at.$$

Maximum speed:

$$\frac{dx}{dt} = ac \cos at; \qquad \text{for } t = 0: \left.\frac{dx}{dt}\right|_{max} = a\cdot c.$$

Average speed:

$$\text{For } at = \frac{\pi}{2}; \qquad x = c \text{ and } \frac{x}{t} = \frac{2ac}{\pi}.$$

The ratio

$$\frac{\left.\dfrac{dx}{dt}\right|_{max}}{\dfrac{x}{t}}$$

is therefore $\dfrac{\pi}{2}$.

the number of letters contained in the successive words of the French verse:

"Que j'aime à faire apprendre
Un nombre utile aux sages.
Immortel Archimède, artiste ingénieur,
Oui, de ton jugement peut priser la valeur."

Translation: "How I like to teach a number, useful to the learned. Immortal Archimedes, skillful investigator, yes, the number can tell the praise of your judgment."

Until recently π had been calculated to 707 decimal places. This figure had been obtained by an Englishman, William Shanks, in 1853. With the help of the modern electronic computing machines, the number of decimal places has now been extended to over 2,000.

The history of the number π dates back 3,500 years, as far as historical records show. The Egyptian Rhind Papyrus, dating back as far as 1700 B.C., gives directions for obtaining the area of a circle. Expressed in modern symbols, its formula with A for the circle's area and d for its diameter, is as follows:

$$A = \left(d - \frac{1}{9}d\right)^2 = d^2\left(1 - \frac{1}{9}\right)^2$$
$$= 4r^2\left(\frac{8}{9}\right)^2 = r^2\frac{4.64}{81} = r^2\frac{256}{81}.$$

The fraction $\frac{256}{81}$, which here takes the place of π, equals in decimals 3.16050 .... Compared with π (3.14159 ...), the difference is 0.01891 ..., or less than $\frac{1}{50}$.

Archimedes expresses π numerically as follows:

$$3\frac{1}{7} > \pi > 3\frac{10}{71}.$$

Expressed in decimals, the same relationship would read

$$3.142857 \ldots > \pi > 3.140845 \ldots .$$

Midway between these two values of Archimedes lies the number 3.141851, which, compared with π is only 0.000259 ... or about $2\frac{1}{2}$ ten-thousandths greater. In ancient China, π was expressed by Ch'ang Höng (A.D. 125) as $\sqrt{10} = 3.162 \ldots$, the accuracy of which is only slightly less than the value given in the Egyptian papyrus. In A.D. 265 Wang Fan expressed the value of π by the fraction $\frac{142}{45}$, or 3.15555 .... In A.D. 470 Ch'ung-chih gave a different fraction: $\frac{355}{113}$, or 3.1415929 ..., which is correct all the way out to 6 decimal places. In India Aryabhata (A.D. 510) expressed π in this way: "Add 4 to 100, multiply by 8 and add 62,000. This is the approximate circumference of a circle whose diameter is 20,000." Thus π appears as the fraction $\frac{62832}{20000}$, which resolves to 3.1416, and is less than one ten-thousandth off.

Though some of these values are sufficiently accurate to have met the practical demands of their times, none reveals any mathematical regularity for the value π. Against the background of the philosophies of antiquity, one can appreciate the great disappointment which this fact caused to mathematicians and philosophers. This failure regarding the outstanding ratio of the most perfect curve to conform to any pattern of mathematical regularity was considered as a blemish upon the divine world order, and never accepted as the ultimate answer.

The anticipations of antiquity regarding π finally proved justified, but the solution was found only as recently as 360 years ago. The value of π was expressed for the first time in a regular mathematical pattern in 1592 by the great French mathematician, François **Viète** (1540–1603), who found

$$\pi = 2 \cdot \frac{1}{\sqrt{\frac{1}{2}} \cdot \sqrt{\frac{1}{2} + \frac{1}{2}\sqrt{\frac{1}{2}}} \cdot \sqrt{\frac{1}{2} + \frac{1}{2}\sqrt{\frac{1}{2} + \frac{1}{2}\sqrt{\frac{1}{2}}}} \ldots}.$$

The denominator is an infinite product of expressions of square roots with a regular structure. The possibility of one such development suggests the possibility of other simpler ones; and actually, in 1655, John **Wallis** (1616–1703), an English mathematician, found

$$\pi = 4 \frac{2 \cdot 4 \cdot 4 \cdot 6 \cdot 6 \cdot 8 \cdot 8 \cdot 10 \cdot 10 \cdot 12 \cdot 12 \ldots}{3 \cdot 3 \cdot 5 \cdot 5 \cdot 7 \cdot 7 \cdot 9 \cdot 9 \cdot 11 \cdot 11 \cdot 13 \ldots}.$$

Here π is expressed by infinite products of numbers, this time in both numerator and denominator of a fraction, but without any roots. In the numerator we find the even numbers, in the denominator the odd numbers. Both appear in pairs with the exception of the first factor in the numerator. Only three years later, in 1658, Viscount Brounecker (1620–1684) expressed the value of π as a continued fraction:

$$\pi = 4 \cdot \cfrac{1}{1 + \cfrac{1^2}{2 + \cfrac{3^2}{2 + \cfrac{5^2}{2 + \cfrac{7^2}{2 + \cfrac{9^2}{2 + \dots}}}}}}$$

which again shows complete regularity, the only varying figures being the squares of the odd numbers.

Progress was on the march. The same century brought the final presentation of π as the limit of an infinite series of the simple fractions made up of the odd numbers as their denominators and with alternating signs, the Leibnitz Series. The regularity which was impossible in decimal expressions of the value of π now became possible through an infinite series of common fractions. Actually, this expression in fractions was more in keeping with the work of the thinkers of antiquity than was that in decimals, which had been in use only since the sixteenth century. That the series was infinite (the transcendence of π was proved by F. Lindemann in 1882) makes the result even more dynamic.

The Leibnitz Series is a fruit of the calculus obtained by one of its inventors. It is derived from expanding the function of arctangent according to Maclaurin's series.

$$f(x) = f(0) + \frac{f'(0)}{1!}\,x + \frac{f''(0)}{2!}\,x^2 + \frac{f'''(0)}{3!}\,x^3 + \dots$$
$$+ \frac{f^n(0)}{n!}\,x^n + \dots .$$

The form for arctan x thus reads:

$$\operatorname{arctan} x = x - \frac{x^3}{3} + \frac{x^5}{5} - \frac{x^7}{7} + \frac{x^9}{9}$$
$$- \frac{x^{11}}{11} + \dots + (-1)^{n-1}\frac{x^{2n-1}}{2n-1} + \dots,$$

which converges for all values of x within the limits

$$-1 \leq x \leq 1.$$

Substituting $x=1$ for an angle of $45°$ (in radians $45° = \pi/4$; tan $45° = 1$) we obtain

$$\frac{\pi}{4} = 1 - \frac{1}{3} + \frac{1}{5} - \frac{1}{7} + \frac{1}{9} - \frac{1}{11} + \frac{1}{13} - \frac{1}{15} + \dots$$

or

$$\pi = 4 \cdot \left(1 - \frac{1}{3} + \frac{1}{5} - \frac{1}{7} + \frac{1}{9} \right.$$
$$\left. - \frac{1}{11} + \frac{1}{13} - \frac{1}{15} + \dots \right).$$

The Leibnitz Series has not been surpassed in all subsequent history in point of its outstanding simplicity. The only later additions were devices for calculating larger numbers of decimals with less effort in the process of computation—in other words, by finding means of developing π through faster convergencies.

By expanding the arcsine in the same way we obtain the formula

$$\operatorname{arcsin} x = x + \frac{1}{2}\cdot\frac{1}{3}x^3 + \frac{1\cdot3}{2\cdot4}\cdot\frac{1}{5}x^5 + \frac{1\cdot3\cdot5}{2\cdot4\cdot6}\cdot\frac{1}{7}x^7$$
$$+ \frac{1\cdot3\cdot5\cdot7}{2\cdot4\cdot6\cdot8}\cdot\frac{1}{9}x^9 + \dots$$

which converges for all values of x within the limits of $-1 \leq x \leq 1$. Substituting $x=1$, we obtain for arcsin 1, corresponding to an angle of $90°$, or, in radians, to $\pi/2$, the formula

$$\frac{\pi}{2} = 1 + \frac{1}{2}\cdot\frac{1}{3} + \frac{1\cdot3}{2\cdot4}\cdot\frac{1}{5} + \frac{1\cdot3\cdot5}{2\cdot4\cdot6}\cdot\frac{1}{7}$$
$$+ \frac{1\cdot3\cdot5\cdot7}{2\cdot4\cdot6\cdot8}\cdot\frac{1}{9} + \dots,$$

a series which, though more complicated than the Leibnitz Series, converges faster. Further series show a still greater convergence—for instance, that which Abraham Sharp used in 1717 to calculate the value of π to 72 decimal places:

$$\pi = 6 \cdot \frac{1}{\sqrt{3}} \cdot \left(1 - \frac{1}{3 \cdot 3} + \frac{1}{3^2 \cdot 5} - \frac{1}{3^3 \cdot 7}\right.$$
$$\left. + \frac{1}{3^4 \cdot 9} - \frac{1}{3^5 \cdot 11} + \ldots \right).$$

To find the value of π geometrically, Deinostratus (350 B.C.) used a curve called the Quadratrix. Its construction is shown in Figure 5. Above and below a horizontal base AB, a quarter of a circle with A as its center and AB as its radius is

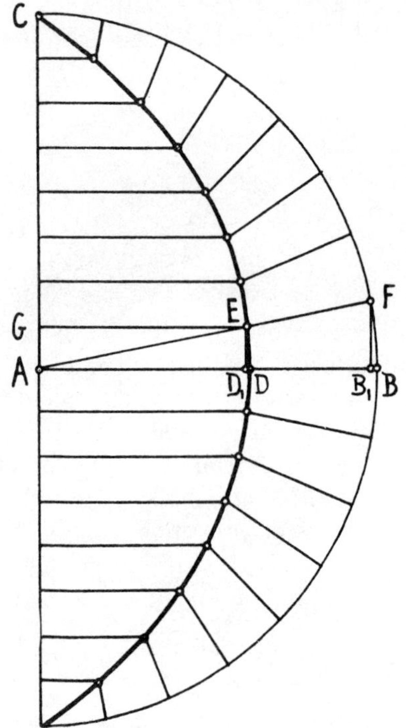

FIG. 5.—GEOMETRIC CONSTRUCTION OF THE VALUE π

drawn and divided into equal parts. In Figure 5 there are eight equal parts above and eight below the base. Then the perpendicular radii are divided into the same numbers of equal parts as the quarters of the circles and through every point of division a horizontal line is drawn. After also adding a radius through each point of division on the circle, we start with the highest point C where the next horizontal line and the next radius intersect, and then continue marking the intersection points of the second horizontal line and the second radius and so forth. The curve

passing through these points is the Quadratrix. Where it cuts the base AB is the point D and the ratio of the line segments AB and AD is

$$\frac{AB}{AD} = \frac{\pi}{2}.^5$$

The geometric aspects of π lead to the famous problem of the quadrature of the circle, the task of constructing a square (quadratum) whose area equals the area of a given circle. The curve in Figure 5 also derives its name from this problem. An outstanding contribution to the quadrature of the circle was made by Archimedes, who found that the area of a circle equals the area of a right triangle, one of whose legs equals the radius and the other the circumference of the circle. This discovery established an equality between the curved area of a circle and the area of a form bounded only by straight lines, and made possible the construction of the quadrature of a circle immediately upon straightening out its circumference. The latter task, so easily performed in actuality

---

[5] The length of the arc BC, being one-quarter of the circumference of a circle, is

$$\frac{2r\pi}{4} = r\frac{\pi}{2}.$$

Its ratio to the radius r is therefore $\frac{\pi}{2}$. The ratio of one-eighth of the arc BC to one-eighth of the radius is therefore also $\frac{\pi}{2}$. The length of the perpendicular from E to AB equals $ED_1 = AG$ which is by construction one-eighth of the radius AC; BF is $\frac{1}{8}$ of BC. Therefore, the ratio BF to ED_1 is still $\frac{\pi}{2}$. What holds good for the eighths holds good for any other fraction. The smaller each part of the arc BC becomes, the closer it approaches the length of the perpendicular FB_1. Through the similarity of the triangles $\triangle AB_1F$ and $\triangle AD_1E$ we obtain the proportion.

$$\frac{AB_1}{AD_1} = \frac{FB_1}{ED_1}.$$

With an increasing number of points of division and the angle FAB decreasing in size, B_1 approaches B, D_1 approaches D, and the ratio $\frac{FB_1}{ED_1}$ the ratio $\frac{FB}{ED^1}$, which equals $\frac{\pi}{2}$.

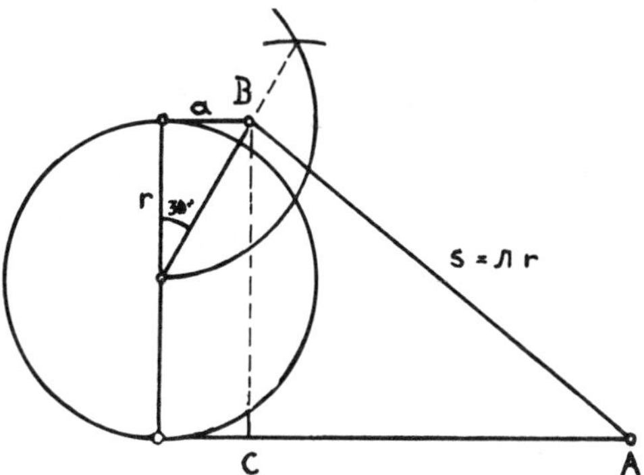

FIG. 6.—CONSTRUCTION BY KOCHANSKY

every time a wheel rolls over a road imprinting on the road its exact circumference with each revolution, has nonetheless been an age-long challenge to masters of geometric construction. Its complete solution is possible only by the use of higher curves. Numerous approximations of this geometric construction have been found, however, which for practical purposes represent a solution. Figure 6 shows the approximation constructed by Kochansky. Through the end-points of the vertical diameter are drawn two tangents to the circle. On each of these tangents a certain point is marked. On the lower tangent this point A is three times the length of the radius of the circle away from the point of tangency, while on the upper the point B is fixed at the intersection of the tangent with the prolonged radius drawn at an angle of 30° to the vertical diameter. The distance AB is then equal to π times the radius.[6] The approximation provides a difference of less than

---

[6] AB computed as the hypotenuse of the right triangle ABC, with its vertical leg $2r$ and its horizontal leg $3r$ minus the distance a, which is one-half the length of the base of an equilateral triangle with the altitude $r \left(a = \dfrac{r}{\sqrt{3}} \right)$ is:

$$AB = \sqrt{(2r)^2 + \left(3r - \frac{r}{\sqrt{3}} \right)^2}$$

$$= r\sqrt{4 + \frac{(3\sqrt{3} - 1)^2}{3}} = r \cdot 3.14153.$$

0.0001, which lies beyond the graphical limit of precision of Figure 6.

With the help of Kochansky's construction, it is possible to effect the quadrature of the circle, as shown in Figure 7, in two steps. In the diagram [above] we recognize Kochansky's construction. The resulting distance is used as the base of a rectangle with an altitude equal to the radius of the circle. According to Archimedes, the area of the circle equals the area of a triangle whose base is the circum

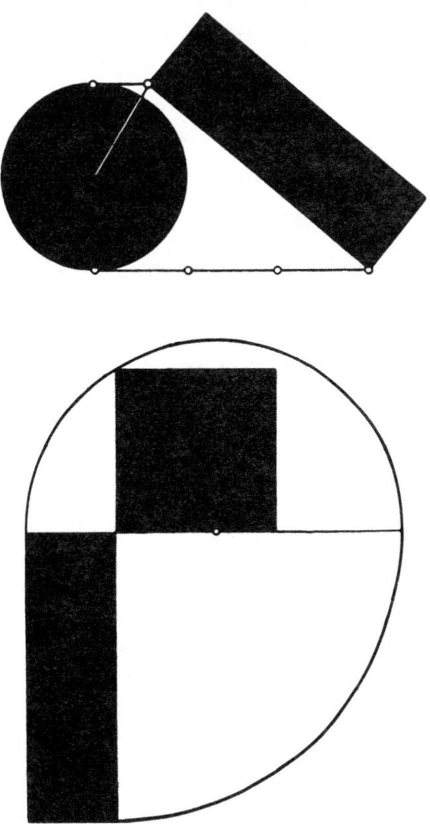

FIG. 7.—CONSTRUCTION OF THE QUADRATURE OF A CIRCLE

ference of the circle and whose altitude is the radius. Therefore, it also equals the rectangle whose base is half the circumference of the circle and whose altitude is the radius. The next step consists of transforming the area of the rectangle into a square—a step accomplished as indicated in Figure 8. The rectangle $ADEF$ is the same as the one in Figure 7. By construction, DB is equal to DE and the intersection of the semicircle above AB with the prolongation of DE determines the point C. $\triangle ABC$ is a right triangle with the altitude h. The area of the square with h as its side equals the area of the rectangle $ADEF$.[7]

The four areas which are marked in black in Figure 7 are equal to one another and show in their sequence the completion of the quadrature of the circle.

Finally, comparing the three great constants of mathematics, G, e, π:

$$G = 0.6180339887 \ldots$$
$$e = 2.7182818284 \ldots$$
$$\pi = 3.1415926535 \ldots$$

in the form of continued fractions:

$$G = \cfrac{1}{1 + \cfrac{1}{1 + \cfrac{1}{1 + \cfrac{1}{1 + \cfrac{1}{1 + \cdots}}}}}$$

$$e = 1 + 1 + \cfrac{1}{1 + \cfrac{1}{2 + \cfrac{1}{1 + \cfrac{1}{1 + \cfrac{1}{4 + \cdots}}}}}$$

$$\pi = 4 \cdot \cfrac{1}{1 + \cfrac{1^2}{2 + \cfrac{3^2}{2 + \cfrac{5^2}{2 + \cfrac{7^2}{2 + \cdots}}}}}$$

G expresses itself through repetitions of the number 1 only, and e through repetitions of 1 and the powers of 2. In the fraction of π, the variable element is the

---

[7] $AD = \dfrac{h}{\tan \angle CAD}$;

$DE = DB = h \cdot \tan \angle BCD$; $\angle BCD = \angle CAD$ (angles whose sides are perpendicular). Therefore

$$AD \cdot DE = \frac{h}{\tan \angle CAD} \cdot h \tan \angle CAD = h^2.$$

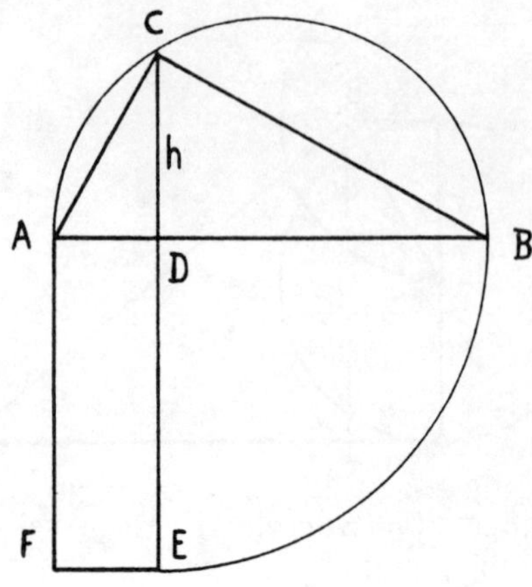

FIGURE 8

series of the squares of the odd numbers.

There is an approximation between π and G which played a major role in the history of the investigations of the proportions of the Great Pyramid in Egypt:

$$\left(\frac{\pi}{4}\right)^2 = 0.6168 \ldots$$
$$G = 0.6180 \ldots .$$

Though the difference between the two values is only 0.0012, their closeness is merely incidental and has no basis in mathematical law. Neither is there any mathematical connection between G and e.

Different is the case with the constants π and e. Between them there is a distinct mathematical relationship. The expansion of e^x, according to Maclaurin's Series is

$$e^x = 1 + x + \frac{x^2}{2!} + \frac{x^3}{3!} + \frac{x^4}{4!} + \frac{x^5}{5!} + \cdots$$
$$+ \frac{x^n}{n!} + \cdots$$

and that of $\sin x$ and that of $\cos x$

$$\sin x = x - \frac{x^3}{3!} + \frac{x^5}{5!} - \frac{x^7}{7!} + \frac{x^9}{9!} + \cdots$$
$$+ (-1)^{n-1} \frac{x^{2n-1}}{(2n-1)!} + \cdots$$
$$\cos x = 1 - \frac{x^2}{2!} + \frac{x^4}{4!} - \frac{x^6}{6!} + \frac{x^8}{8!} + \cdots$$
$$+ (-1)^{n-1} \frac{x^{2n-2}}{(2n-2)!} + \cdots .$$

The two series for $\sin x$ and $\cos x$ together furnish all the terms of the series of e^x, with the only discrepancy in their signs, a difficulty which does not exist for the hyperbolic sines and cosines. Their expansions have only positive terms:

$$\sinh x = x + \frac{x^3}{3!} + \frac{x^5}{5!} + \frac{x^7}{7!} + \ldots + \frac{x^{2n-1}}{(2n-1)!}$$

$$\cosh x = 1 + \frac{x^2}{2!} + \frac{x^4}{4!} + \frac{x^6}{6!} + \ldots + \frac{x^{2n-2}}{(2n-2)!}.$$

Therefore it readily appears that $e^x = \sinh x + \cosh x$. An analogous result for the trigonometric functions can be obtained if we substitute for x its product with the imaginary unit:

$$e^{ix} = 1 + ix - \frac{x^2}{2!} - i\frac{x^3}{3!} + \frac{x^4}{4!} + i\frac{x^5}{5!}$$
$$- \frac{x^6}{6!} - i\frac{x^7}{7!} + \ldots$$

and separate the real and the imaginary terms. Thus, the result is

$$e^{ix} = 1 - \frac{x^2}{2!} + \frac{x^4}{4!} - \frac{x^6}{6!} + \ldots$$
$$+ i\left(x - \frac{x^3}{3!} + \frac{x^5}{5!} - \frac{x^7}{7!} + \ldots \right)$$

or

$$e^{ix} = \cos x + i \sin x,$$

a formula which finds wide application in the solving of differential equations, particularly of those connected with all types of vibrations. Substituting $x = \pi$, we obtain

$$e^{i\pi} = \cos \pi + i \sin \pi = -1 + i \cdot 0 = -1,$$

which results in the formula,

$$e^{\pi i} = -1.$$

It is this formula which prompted David Eugene Smith to use it in the mathematical credo placed in his library:

The Science Venerable:

Voltaire once remarked—"One merit of poetry few will deny; it says more and in fewer words than prose." With equal significance we may say, "One merit of mathematics few will deny; it says more and in fewer words than any other science." The formula $e^{\pi i} = -1$ expresses a world of thought, of truth, of poetry, and of religious spirit, for "God eternally geometrizes."

BIBLIOGRAPHY

BALL, W. W. ROUSE. *Mathematical Recreations and Essays*, revised by H. S. M. Coxeter. New York: The Macmillan Co., 1939.

BARAVALLE, HERMANN VON. *Die Geometrie des Pentagramms und der Goldene Schnitt.* Stuttgart: J. Ch. Mellinger Verlag, 1950.

BINDEL, ERNST. *Die Aegyptischen Pyramiden.* Stuttgart: Verlag Freie Waldorfschule GMBH, 1932.

DE MORGAN, AUGUSTUS. *A Budget of Paradoxes.* Volume II. Chicago and London: The Open Court Publishing Co., 1915.

GHYKA, MATILA C. *Esthétique des Proportions dans la Nature et dans les Arts.* Paris: Librairie Gallimard, 1927.

GRANVILLE, W. A., SMITH, P. F., and LONGLEY, W. R. *Elements of the Differential and Integral Calculus.* Boston: Ginn & Co., 1934.

KASNER, EDWARD, and NEWMAN, JAMES. *Mathematics and the Imagination.* New York: Simon & Schuster, 1940.

ROSENBERG, KARL. *Das Rätsel der Cheopspyramide*, Band 154. Wien: Deutsche Hausbücherei, Österreichischer Bundes Verlag, 1925.

SMITH, DAVID EUGENE. *History of Mathematics.* Boston: Ginn & Co., 1925.

SOME COMMENTS on e

By NORMAN SCHAUMBERGER

Bronx Community College
Bronx, New York

THE number represented by e is most frequently defined as the limit of the sequence

$$S_n = \left(1 + \frac{1}{n}\right)^n$$

as $n \to \infty$. In many introductory calculus texts either e is simply defined in this way without much further discussion, or a plausible but incorrect proof is presented that asserts that the sequence

$$S_n = \left(1 + \frac{1}{n}\right)^n$$

does approach a limit as $n \to \infty$. A simple but valid proof can be given, based on the following statements:

(1) $S_n = \left(1 + \frac{1}{n}\right)^n$ is an increasing sequence; that is, $S_1 < S_2 < \cdots < S_n < S_{n+1} < \cdots$.

(2) $T_n = \left(1 + \frac{1}{n}\right)^{n+1}$ is a decreasing sequence; that is, $T_1 > T_2 > \cdots > T_n > T_{n+1} > \cdots$.

Assume that (1) and (2) are true. It is a simple matter to show that S_n and T_n both tend to one and the same limit. For by (1) the sequence S_n is increasing. Moreover, since

$$T_n = \left(1 + \frac{1}{n}\right)S_n,$$

then $S_n < T_n$, and since by (2) T_n is decreasing, we have

$$S_n < T_n \leqq T_1 = \left(1 + \frac{1}{1}\right)^2 = 4.$$

Thus S_n is an increasing sequence that is bounded above by 4 and so must approach a limit. Similarly, using the fact that S_n is increasing with n, we have

$$S_n \geqq S_1 = \left(1 + \frac{1}{1}\right)^1 = 2.$$

Since $T_n > S_n$, it follows that T_n is a decreasing sequence that is bounded below by 2. Thus T_n tends to a limit. Furthermore, since

$$\lim_{n\to\infty} \frac{T_n}{S_n} = \lim_{n\to\infty} \left(1 + \frac{1}{n}\right) = 1,$$

the limits of S_n and T_n are identical. We call this common limit the number e. Hence with the assumptions that (1) and (2) are true, it is quite easy to show that

$$\lim_{n\to\infty} S_n = \lim_{n\to\infty} T_n = e.$$

Furthermore, it follows that for all positive integers n, $S_n < e < T_n$. For example, if $n = 200$, we have

$$2.711 < \left(1 + \frac{1}{200}\right)^{200} = S_{200} < e < T_{200}$$

$$= \left(1 + \frac{1}{200}\right)^{201} < 2.725.$$

Thus from (1) and (2) we are able to show not only that the number e exists as a limit but also that the proof of existence can actually yield the value of e to as many decimal places as desired, though the method is quite cumbersome.

It remains to verify (1) and (2). The usual proofs of these statements either employ the binomial theorem (see Shklarsky, Chentzov, and Yaglom 1962, pp.

246–47, for example) or are based on the classical inequality that states that the arithmetic mean of a set of n positive numbers is greater than or equal to their geometric mean (see Mendelsohn 1951, p. 563). The first approach is somewhat messy, and the second uses the statement of a theorem that is rarely proved in elementary calculus. Consider the alternative proofs of (1) and (2) that require only mathematics frequently taught in high school.

To establish (1) it is sufficient to show that for each positive integer n,

$$\left(1 + \frac{1}{n+1}\right)^{n+1} > \left(1 + \frac{1}{n}\right)^{n}.$$

Since $1 + \frac{1}{n} > 1$ and $\frac{j}{n+1} < \frac{n}{n+1}$ when $j = 0, 1, 2, \cdots, n-1$, it follows that

$$\left(1 + \frac{1}{n}\right)^{j/(n+1)} < \left(1 + \frac{1}{n}\right)^{n/(n+1)}$$

when $j = 0, 1, 2, \ldots, n-1$. Consequently, we get

$$\sum_{j=0}^{n-1}\left(1 + \frac{1}{n}\right)^{j/(n+1)} < \sum_{j=0}^{n-1}\left(1 + \frac{1}{n}\right)^{n/(n+1)}$$

$$= n\left(1 + \frac{1}{n}\right)^{n/(n+1)}.$$

Summing the geometric series on the left gives

$$\frac{\left(1 + \frac{1}{n}\right)^{n/(n+1)} - 1}{\left(1 + \frac{1}{n}\right)^{1/(n+1)} - 1} < n\left(1 + \frac{1}{n}\right)^{n/(n+1)}$$

or

$$\left(1 + \frac{1}{n}\right)^{n/(n+1)} - 1$$

$$< n\left(1 + \frac{1}{n}\right) - n\left(1 + \frac{1}{n}\right)^{n/(n+1)},$$

and therefore

$$(n + 1)\left(1 + \frac{1}{n}\right)^{n/(n+1)} < (n + 1) + 1.$$

Thus

$$\left(1 + \frac{1}{n}\right)^{n/(n+1)} < \left(1 + \frac{1}{n+1}\right),$$

and raising both sides to the $(n + 1)$st power, we have

$$\left(1 + \frac{1}{n}\right)^{n} < \left(1 + \frac{1}{n+1}\right)^{n+1}.$$

Proof of (2). This is equivalent to showing that

$$\left(1 + \frac{1}{n-1}\right)^{n} > \left(1 + \frac{1}{n}\right)^{n+1}$$

when n is a positive integer.

Since $1 - \frac{1}{n} < 1$, it follows that

$$\left(1 - \frac{1}{n}\right)^{j/(n+1)} > \left(1 - \frac{1}{n}\right)^{n/(n+1)}$$

when $j = 0, 1, 2, \ldots, n-1$. Consequently,

$$\sum_{j=0}^{n-1}\left(1 - \frac{1}{n}\right)^{j/(n+1)} > \sum_{j=0}^{n-1}\left(1 - \frac{1}{n}\right)^{n/(n+1)}$$

$$= n\left(1 - \frac{1}{n}\right)^{n/(n+1)}.$$

Summing the geometric series on the left, we get

$$\frac{1 - \left(1 - \frac{1}{n}\right)^{n/(n+1)}}{1 - \left(1 - \frac{1}{n}\right)^{1/(n+1)}} > n\left(1 - \frac{1}{n}\right)^{n/(n+1)}.$$

This can be written as

$$n > (n + 1)\left(1 - \frac{1}{n}\right)^{n/(n+1)}$$

or

$$\frac{n}{n+1} > \left(\frac{n-1}{n}\right)^{n/(n+1)},$$

and raising both sides to the $(n + 1)$st power, we obtain

$$\left(\frac{n}{n+1}\right)^{n+1} > \left(\frac{n-1}{n}\right)^{n}.$$

Inverting both sides reverses the inequality, and we have

$$\left(\frac{n+1}{n}\right)^{n+1} < \left(\frac{n}{n-1}\right)^{n}.$$

Thus

$$\left(1 + \frac{1}{n}\right)^{n+1} < \left(1 + \frac{1}{n-1}\right)^{n},$$

which is what we wished to prove.

REFERENCES

Kazarinoff, N. D. *Analytic Inequalities*, pp. 40–42. New York: Holt, Rinehart & Winston, 1961.

Mendelsohn, N. S. "An Application of a Famous Inequality." *American Mathematical Monthly* 58 (1951).

Shklarsky, D. O., N. N. Chentzov, and I. M. Yaglom. *The USSR Olympiad Problem Book.* San Francisco: W. H. Freeman & Co., 1962.

Bibliography: *Special Numbers—*
e and π

Sherk, Wilfred H. Approximate Values of π. 2 (March 1910): 87–93.
   Remarks on approximating π by such numerical methods as Cotes's method, Simpson's one-third rule, Simpson's three-eighths rule, and Weddle's rule.

Jones, Phillip S. What's New about π? Notes on Older Facts. 43 (March 1950): 120–22.
   Some historical remarks concerning decimal approximations to π.

Carnahan, Walter H. Pi and Probability. 46 (February 1953): 65–66, 70.
   Remarks on Buffon's needle problem as well as coin tossing on a cross-ruled board.

Jones, Phillip S. Recent Discoveries in Babylonian Mathematics 1: Zero, Pi, and Polygons. 50 (February 1957): 162–65.
   Historical remarks dealing with Babylonian mathematical achievements.

Eves, Howard. The Latest about π. 55 (February 1962): 129–30.
   Remarks on some computer calculations of π.

Thumm, Walter. Buffon's Needle: Stochastic Determination of π. 58 (November 1965): 601–7.
   Detailed remarks on Buffon's needle problem.

Read, Cecil B. Shanks, Pi, and Coincidence. 60 (November 1967): 761–62.
   Some historical remarks concerning decimal approximations to π.

Te Selle, David W. Pi, Polygons, and a Computer. 63 (February 1970): 128–32.
   Remarks on estimating π from perimeters of regular polygons inscribed in, and circumscribed about, a circle. Includes a BASIC program for the calculation.

Smithson, Thomas W. An Eulerian Development for Pi: A Research Project for High School Students. 63 (November 1970): 597–608.
   Historical remarks on infinite series expansions for π. Includes a CDC 6400 program for the series summation.

Oliver, Bernard M. The Key of *e*. 65 (January 1972): 5–8.
   Remarks concerning the definition of *e* and its significance to the study of trigonometric functions.

Einhorn, Erwin. A Method for Approximating the Value of π with a Computer Application. 66 (May 1973): 427–30.
   Remarks on estimating π using circular areas. Includes a BASIC program for the calculation.

Hatcher, Robert S. Some Little-Known Recipes for π. 66 (May 1973): 470–74.
   Includes various series expansions for π.

11

General Applications

Calculus teachers search continually for simple yet nontrivial applications of basic concepts. Specific applications of differentiation or integration are to be found in other sections of this book. Two general applications are described here.

The first of these, discussed by Myers, is the speedometer, which can be used to illustrate the processes of both differentiation and integration. The second is a timely problem in space technology, presented by Stretton —the calculation of escape velocity.

Discussing applications reinforces the learning process. Students need to understand not only what they are learning but also its relevance to real life.

The Meaning of Derivative and Integral Illustrated by the Speedometer

By SHELDON S. MYERS

The common, everyday device known as the speedometer constantly illustrates the derivative and integral of calculus. The teaching of an elective course in calculus for the twelfth grade was recommended by the National Committee of 1923. One of the ardent proponents of the teaching of calculus in high school was the late John A. Swenson who wrote the chapter, "Selected Topics in Calculus for the High School," in the Third Yearbook of The National Council of Teachers of Mathematics. The many teachers today who are teaching the calculus to seniors are anxious to find simple ways to illustrate the meaning of derivative and integral within the experience of students. The following example occurred to me and has proved successful in clarifying the meaning of derivative and integral for high school seniors.

The speedometer is a very familiar instrument which illustrates both the derivative and the integral. The part of the instrument which shows the speed at any given moment is really giving the derivative continuously of the curve plotting cumulative distance against the time for this particular automobile. The part of the instrument which records the total trip is really giving the integral of the curve plotting speed against time for this particular automobile. These two curves with the derivative of one and the integral of the other can be compared by the graphs below:

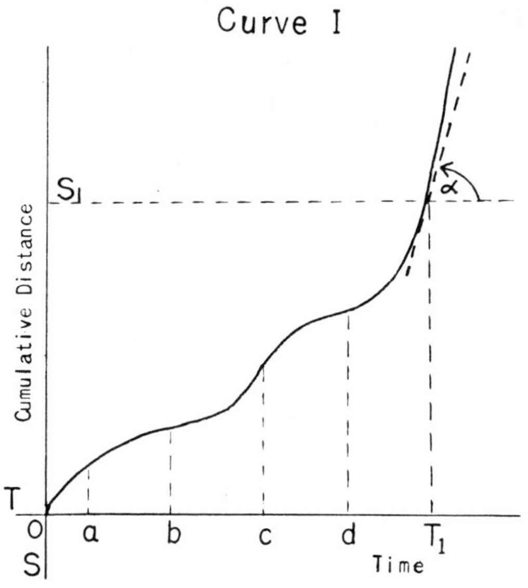

Ordinate is Total Trip.

If $S = f(T)$, then

$\frac{dS}{dT}$ = derivative of $f(T)$

$= f_1(T) = R.$

If $T = T_1$, then $\frac{dS}{dT} = R_1 = f_1(T_1)$

or $\tan\alpha = R_1.$

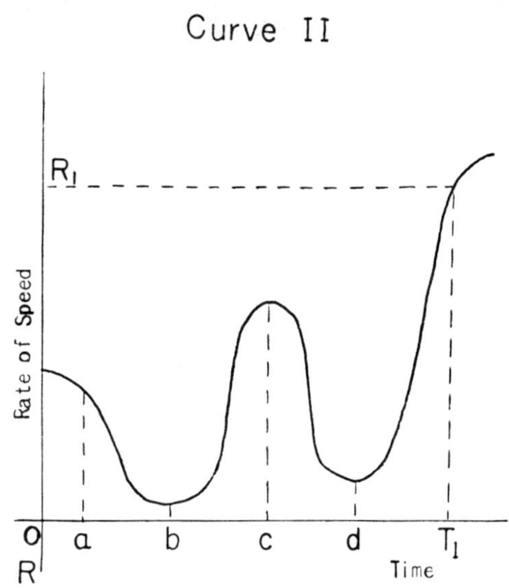

Ordinate is Speed.

If $R = f_2(T)$, then

$$\int_0^{T_1} R\, dT = \int_0^{T_1} f_2(T)\, dT = S_1$$

$= $ total trip.

$S_1 = $ total trip = area under

curve between $T = 0$ and $T = T_1.$

These two curves are interesting in that the integral of Curve II is the ordinate of Curve I, while the derivative of Curve I is the ordinate of Curve II. Furthermore, the linear portions of Curve I are the horizontal portions of Curve II. The steep portions of Curve I are high portions of Curve II, while the more horizontal portions of Curve I are the lower portions of Curve II. Check this by comparing the corresponding times a, b, c, d, and T_1 on both curves. Although this illustration may not be appropriate at the very beginning of the study of calculus, there is a stage, to be determined by the situation, when this illustration will greatly strengthen understanding.

The velocity of escape

WILLIAM C. STRETTON, *Lyons Township High School,*
La Grange, Illinois.

Physics for the modern world in the mathematics class

IN THESE DAYS of artificial satellites and moon shots, one often hears reference to the velocity of escape, the velocity a projectile must have in order to leave the earth and not return. It is not too difficult to derive the velocity of escape using known laws of physics, elementary mathematics, and a plausible assumption that can be verified by calculus.

We start with the assumption that the work necessary to remove a body of mass m from the earth's surface to infinity is given by

$$W = \frac{GMm}{s_0} \qquad (1)$$

where G is the gravitational constant, M the mass of the earth, and s_0 the radius of the earth. We can make this assumption plausible by using Newton's law of gravitation and some elementary mathematics. According to Newton's law, the force of attraction between two bodies of mass M and m respectively is given by

$$\frac{GMm}{s^2}$$

where s is the distance between their centroids. Consider mass M at 0 and mass m at A (Fig. 1). What work is necessary to move m from A to B? Now work is given by force times distance. The force at A is

$$F_A = \frac{GMm}{\overline{OA}^2}$$

and the force at B is

$$F_B = \frac{GMm}{\overline{OB}^2} .$$

Since $\overline{OA} < \overline{OB}$ it follows that $\overline{OA}^2 < \overline{OA} \cdot \overline{OB} < \overline{OB}^2$. Hence

$$F = \frac{GMm}{\overline{OA} \cdot \overline{OB}}$$

is an approximation for the average force between A and B, the approximation becoming better as $\overline{AB}$ becomes smaller. If $\overline{AB}$ is small, then, the work necessary to move m from A to B is very nearly equal to

$$\frac{GMm}{\overline{OA} \cdot \overline{OB}} (\overline{OB} - \overline{OA}) = \frac{GMm}{\overline{OA}} - \frac{GMm}{\overline{OB}} .$$

Similarly, the work in going from B to C is

$$\frac{GMm}{\overline{OB}} - \frac{GMm}{\overline{OC}},$$

from C to D,

$$\frac{GMm}{\overline{OC}} - \frac{GMm}{\overline{OD}},$$

etc. On adding we see that the total work necessary to move m from A to D is approximately

$$\frac{GMm}{\overline{OA}} - \frac{GMm}{\overline{OD}} .$$

If now we increase the number of points

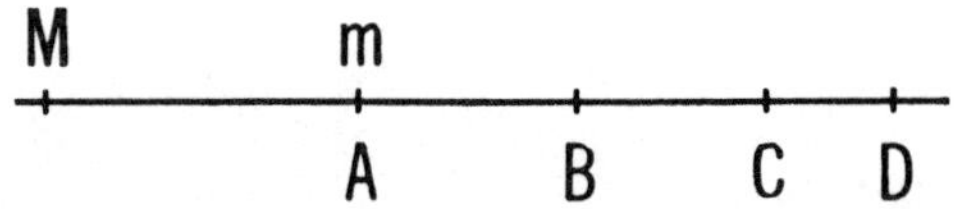

Figure 1

$B, C, \cdots$ between A and D in such a way as to make the segments $AB, BC, \cdots$ all extremely small, the accuracy of this approximation is greatly improved. Since the formula however, remains the same in this process, it evidently must be exact. From this we see that the work necessary to move a mass m from the surface of the earth to a point whose distance is s_1 from the center would be

$$\frac{GMm}{s_0} - \frac{GMm}{s_1} .$$

If $s_1 \to \infty$, then

$$\frac{GMm}{s_1} \to 0$$

and the work needed to take m to infinity is seen to be

$$\frac{GMm}{s_0} .$$

Approximations were used in deriving the above result. For students with a calculus background, the result can be rigorously derived by use of the fundamental theorem and the evaluation of the resulting definite integral,

$$\lim_{s \to \infty} \int_{s_0}^{s} \frac{GMm}{x^2} dx.$$

We now make use of another known law from physics, the expression for kinetic energy. It tells us that the work necessary to give a body of mass m a velocity v is given by $\frac{1}{2}mv^2$. It follows that the work necessary to remove a body of mass m to infinity,

$$\frac{GMm}{s_0} ,$$

must equal $\frac{1}{2}mv_0^2$ where v_0 is the initial velocity. Equating these and solving for v_0 we get

$$v_0 = \sqrt{\frac{2GM}{s_0}} , \tag{2}$$

the velocity of escape. We can put this in a more convenient form for computation by making a substitution based on what follows. Newton's second law tells us that force is equal to mass times acceleration. At the surface of the earth the acceleration due to gravity is denoted by g and is approximately 32 ft. per sec. per sec. Hence at the earth's surface we must have

$$mg = \frac{GMm}{s_0^2} .$$

From this we get

$$\frac{GM}{s_0} = s_0 g .$$

Substituting this result in (2) we obtain

$$v_0 = \sqrt{2 s_0 g} . \tag{3}$$

Using $s_0 = 4,000$ miles, $g = 32$ ft. per sec. per sec., we can calculate v_0 to be about 36,000 ft. per sec. or 7 mi. per sec.

We can use an entirely different approach with students who have a calculus background. This time, again, we use the law of gravitation which can be stated thus: the acceleration due to gravity at any point above the earth's surface varies inversely as the square of the distance from the center of the earth. Using the notation of calculus, this can be written

$$\frac{dv}{dt} = -\frac{s_0^2}{s^2} g, \tag{4}$$

the negative sign appearing because we are considering velocity directed upward whereas the acceleration is directed downward. Using the chain rule, we can then write

$$\frac{dv}{dt} = \frac{dv}{ds} \cdot \frac{ds}{dt} = v \frac{dv}{ds} .$$

On substituting this result in (4) we have

$$v \frac{dv}{ds} = -\frac{s_0^2}{s^2} g. \tag{5}$$

This simple differential equation can be solved quite easily yielding

$$\frac{v^2}{2}=\frac{s_0^2 g}{s}+c. \tag{6}$$

The constant can be determined from initial conditions for when $v=v_0$, $s=s_0$ and hence

$$c=\frac{v_0^2}{2}-s_0 g.$$

Substituting this value of c in (6) we then have

$$\frac{v^2}{2}=\frac{s_0^2}{s}\,g+\frac{v_0^2}{2}-s_0 g, \tag{7}$$

a relation between the velocity v of a projectile at any height s above the earth's surface if given initial velocity v_0.

Suppose, now, we wish to fire a projectile to a height s_1 and have it return. This means that at $s=s_1$ in (7) we must have $v=0$. Substituting in (7) and solving for v_0 we have

$$v_0=\sqrt{2s_0 g\left(1-\frac{s_0}{s_1}\right)} \tag{8}$$

a relation between the initial velocity v_0 and the height s_1 reached by a projectile. If now we wish to fire the projectile with such velocity that it will not return, we simply require that $s_1\to\infty$. As

$$s_1\to\infty, \quad \frac{s_0}{s_1}\to 0,$$

and (8) becomes

$$v_0=\sqrt{2s_0 g} \tag{9}$$

which agrees with (3) above.

It is instructive to take another look at (8). The quantity in parentheses can be written

$$\frac{s_1-s_0}{s_1}.$$

If s_1 is only slightly larger than s_0 then s_0/s_1 will be approximately equal to unity. Then, if we let $h=s_1-s_0$, (8) becomes

$$v_0=\sqrt{2gh}, \tag{10}$$

a familiar formula from elementary physics texts. We see that (10) is only approximately true but nevertheless useful provided h is small compared to s_0, that is, provided the height is small enough so that the acceleration due to gravity is not measurably different from g, the acceleration due to gravity at the earth's surface.

Bibliography: *General Applications*

Chevrolet Motor Division. The Mathematics of the Automobile. 31 (May 1938): 209–15.
Remarks on the application of mathematics and physics to automotive engineering.

Schaaf, William L. Mathematics and Billiards. 50 (May 1957): 384–85.
A summary of the equations of motion occurring in a mathematical analysis of billiards.

Author Index

Page numbers in italic type refer to authors cited in bibliographies.

Subject Index

FUNDERBURG LIBRARY

MANCHESTER COLLEGE

515.08
C126n

WITHDRAWN
from
Funderburg Library